SpringerBriefs in Applied Sciences and Technology

SpringerBriefs present concise summaries of cutting-edge research and practical applications across a wide spectrum of fields. Featuring compact volumes of 50 to 125 pages, the series covers a range of content from professional to academic.

Typical publications can be:

- A timely report of state-of-the art methods
- An introduction to or a manual for the application of mathematical or computer techniques
- A bridge between new research results, as published in journal articles
- A snapshot of a hot or emerging topic
- An in-depth case study
- A presentation of core concepts that students must understand in order to make independent contributions

SpringerBriefs are characterized by fast, global electronic dissemination, standard publishing contracts, standardized manuscript preparation and formatting guidelines, and expedited production schedules.

On the one hand, **SpringerBriefs in Applied Sciences and Technology** are devoted to the publication of fundamentals and applications within the different classical engineering disciplines as well as in interdisciplinary fields that recently emerged between these areas. On the other hand, as the boundary separating fundamental research and applied technology is more and more dissolving, this series is particularly open to trans-disciplinary topics between fundamental science and engineering.

Indexed by EI-Compendex, SCOPUS and Springerlink.

Shafiq Bin Suhaimi · Solehuddin Shuib ·
Hamid Yusoff

Flapping Wing Micro Air Vehicles

Bio-Inspired Aerodynamics

Shafiq Bin Suhaimi
School of Mechanical Engineering
College of Engineering
Universiti Teknologi MARA
Shah Alam, Selangor, Malaysia

Solehuddin Shuib
School of Mechanical Engineering
College of Engineering
Universiti Teknologi MARA
Shah Alam, Selangor, Malaysia

Hamid Yusoff
School of Mechanical Engineering
College of Engineering
Universiti Teknologi MARA Cawangan Pulau Pinang
Permatang Pauh, Pulau Pinang, Malaysia

ISSN 2191-530X ISSN 2191-5318 (electronic)
SpringerBriefs in Applied Sciences and Technology
ISBN 978-981-96-2907-7 ISBN 978-981-96-2908-4 (eBook)
https://doi.org/10.1007/978-981-96-2908-4

© The Editor(s) (if applicable) and The Author(s), under exclusive license to Springer Nature Singapore Pte Ltd. 2025

This work is subject to copyright. All rights are solely and exclusively licensed by the Publisher, whether the whole or part of the material is concerned, specifically the rights of translation, reprinting, reuse of illustrations, recitation, broadcasting, reproduction on microfilms or in any other physical way, and transmission or information storage and retrieval, electronic adaptation, computer software, or by similar or dissimilar methodology now known or hereafter developed.
The use of general descriptive names, registered names, trademarks, service marks, etc. in this publication does not imply, even in the absence of a specific statement, that such names are exempt from the relevant protective laws and regulations and therefore free for general use.
The publisher, the authors and the editors are safe to assume that the advice and information in this book are believed to be true and accurate at the date of publication. Neither the publisher nor the authors or the editors give a warranty, expressed or implied, with respect to the material contained herein or for any errors or omissions that may have been made. The publisher remains neutral with regard to jurisdictional claims in published maps and institutional affiliations.

This Springer imprint is published by the registered company Springer Nature Singapore Pte Ltd.
The registered company address is: 152 Beach Road, #21-01/04 Gateway East, Singapore 189721, Singapore

If disposing of this product, please recycle the paper.

Competing Interests The authors have no competing interests to declare that are relevant to the content of this manuscript.

About This Book

This book focuses on bio-inspiration design focuses on the design of a wing for micro air vehicles. The book looks into the examples of the design method and applying a novel bat inspired wing design and testing the design using both experimental and simulation approach. In terms of looking at examples, the book studies all of the works done that uses bio-inspiration approach and breaking down its different aspects such as wing type, how the mimicry was done, and the results that arrived from the works that have been done. From this, a gap was identified and is then used to form a new wing design. This books offers an interesting method for bio-inspiration design that answers one of the main problem with taking forms in nature for mechanical design which is the question of how to deal with the complexity in nature where it cannot be easily replicated in manmade machines.

Contents

1 **Introduction** 1
   1.1   Introduction to Bio-inspired Design 1
   References 5

2 **Bio-inspired Flyers** 7
   2.1   Introduction 7
   2.2   Natural Bats 8
   2.3   Bat Wing Morphology 8
   2.4   Bat Flight Behaviour 10
   2.5   Bat Flight Kinematics 12
   2.6   Previous Studies 14
   2.7   Bat Wing Geometry Mimicry 15
   2.8   Wing Kinematics Mimicry 20
   2.9   Wing Aerodynamics 28
   2.10 Knowledge Gap 32
   References 34

3 **Bat-Inspired Wing Design** 39
   References 52

4 **Bat-Inspired Wing Performance** 53
   Reference 71

5 **Final Thoughts** 73

Index 77

Abbreviations

AoA	Angle of Attack
CAD	Computer Aided Design
CFD	Computational Fluid Dynamics
DAQ	Data Acquisition System
FEA	Finite Element Analysis
FSI	Fluid-Structure Interactions
FW-MAV	Flapping Wing Micro Air Vehicles
LEV	Leading Edge Vortex
MAV	Micro Air Vehicles
RANS	Reynolds Averaged Navier Stokes
TRV	Trailing Edge Vortex
WRV	Wing Root Vortex
WTV	Wing Tip Vortex

Symbols

ρ Density of air
ω Wing angular velocity
ω_N Cut off frequency
ε Strain

List of Figures

Fig. 2.1	Anatomical parts of a natural wing (Sterbing-D'Angelo et al. 2011)	9
Fig. 2.2	Motions of a batwing (Yu and Guan 2015)	13
Fig. 2.3	Wing models with skeletal features and membrane wings (Bahlman et al. 2013)	17
Fig. 2.4	Wing models that use conventional or generic wing shapes (Heathcote et al. 2008)	18
Fig. 2.5	2D wing models (Geissler and van der Wall 2017)	20
Fig. 2.6	Wing motion model based on natural wing motion tracking (Tian et al. 2006)	22
Fig. 2.7	Wing motion model that combines several wing motion types (Bahlman et al. 2014)	23
Fig. 2.8	Wing motion model that uses pure flapping motion (Huera-Huarte and Gharib 2017)	25
Fig. 2.9	Fixed-wing motion model (Wang et al. 2017)	27
Fig. 2.10	Previous study matrix	33
Fig. 3.1	Flow chart of the wing geometry generation process	40
Fig. 3.2	Final wing model	42
Fig. 3.3	Flapper used in the study	46
Fig. 3.4	Fabricated flapper using 3D-printing	47
Fig. 3.5	The sub-sonic open channel wind tunnel	48
Fig. 4.1	Comparison results for the $C_{L\,avg}$ against AoA	55
Fig. 4.2	Comparison results for the $C_{D\,avg}$ against AoA	56
Fig. 4.3	Comparison results for the $C_{L\,avg}/C_{D\,avg}$ against AoA	56
Fig. 4.4	Simulation and test results comparison where; **a** is for the 99 vertices wing, **b** is for the 49 vertices wing, **c** is for the 29 vertices wing, **d** is for the 21 vertices wing, and **e** is for the 5 vertices wing	58
Fig. 4.5	Location of the cross-section plane	61
Fig. 4.6	Example of the formation of LEV	61
Fig. 4.7	LEV development for 99 vertices wing	62

Fig. 4.8	LEV development for different AoA for 99 vertices wing at 0.25 T time phase	62
Fig. 4.9	Close up of the TEV	63
Fig. 4.10	LEV for all wing geometries at 30° AoA at 0.25 T	64
Fig. 4.11	Trailing edge spillover where; **a** is the 99 vertices wing at 30° AoA at 0.25 T and **b** is the 5 vertices wing at 30° AoA at 0.25 T	66
Fig. 4.12	Wingtip vortex visualisation cross-section plane	66
Fig. 4.13	Example of the formation of WTV	67
Fig. 4.14	WTV development for 99 vertices wing at 0° AoA	68
Fig. 4.15	WTV development for 99 vertices wing downstroke motion	69
Fig. 4.16	WTV development for 99 vertices wing upstroke motion	70

List of Tables

Table 2.1	List of previous works done in 3d realistic batwing models	16
Table 2.2	List of previous works done in wing models with skeletal features	17
Table 2.3	List of previous works done with the margin shape model	18
Table 2.4	List of previous works done with conventional or generic wing shapes	19
Table 2.5	List of previous works done in 2D wing models	21
Table 2.6	List of previous works done with natural wing motion tracking	22
Table 2.7	List of previous works done with flapping motion combined with several motion types	23
Table 2.8	List of previous works done with flapping motion combined with one other wing motion types	24
Table 2.9	List of previous works done with a pure flapping motion	26
Table 2.10	List of previous works done with fixed-wing	27
Table 3.1	Final wing model dimensions	42
Table 3.2	List of all generated wings	44
Table 3.3	Flight conditions value	45
Table 3.4	Mechanical properties of PLA	47

Chapter 1
Introduction

Abstract This chapter introduces to the concept of bio-inspiration design, and basics concepts of micro air vehicles. This chapter also showcase the design problems ins MAV designs and the problems of bio-inspiration approach.

1.1 Introduction to Bio-inspired Design

This book focuses on two main aspects; the first aspect focuses on the technical problem involved with Micro Air Vehicles or MAVs where, due to the nature of MAVs themselves, the lift generation and aerodynamics efficiency become an issue. The second aspect of this book involves the design problem of bio-inspired flapping-wing design where there is a lack of methods for translating the observations found in nature to mechanical applications. These two aspects; the aerodynamics problem and the bio-inspiration design problem will be the main topics in this book where every part of the book will lead to solving one or both two aspects.

Micro Air Vehicles (MAVs) can be defined as an aerial vehicle that has a wingspan of less than 15 cm and a take-off weight of less than 200 g (Pines and Bohorquez 2006). The main advantages that MAVs have above all other manned or unmanned aerial vehicles lie in their small size where MAVs can operate in small and tight spaces, where surveillance can be made in hard-to-reach areas, and where humans and other types of equipment are unable to make. The other advantage of the small size of a MAV is its ability to fly unnoticed by others where the small size of the MAV can fly at a certain distance away without causing detection. This means that MAVs are used in military, law enforcement, and industrial applications. MAVs are useful for surveillance work in a hostile or hazardous area where direct human contact might not be desirable. Some of the examples of the potential application of a MAV include pollution inspection, pipe inspection, and deadly indoor inspection for industrial application, crowd surveillance, hostile reconnaissance, and traffic accident reporting for military or law enforcement applications. There are three main categories of MAV wing types: fixed wings, rotary wings, and flapping (or bio-inspired) wings (Galinski and Zbikowski 2007).

© The Author(s), under exclusive license to Springer Nature Singapore Pte Ltd. 2025
S. B. Suhaimi et al., *Flapping Wing Micro Air Vehicles*,
SpringerBriefs in Applied Sciences and Technology,
https://doi.org/10.1007/978-981-96-2908-4_1

The first type of MAV wing is the fixed wing where it is known to have the best forward flight capabilities with the fastest forward flight speed and the best endurance. However, fixed wings have their drawback in terms of manoeuvrability since these require constant forward speed for lift and are unable to achieve tighter turns. The second type of MAV wing is the rotary wing which, unlike fixed wings, can achieve tighter turns and can hover in place allowing for navigations in small places and prolonged surveillance of a single fixed position. However, rotary wings suffer from inefficiency.

This leaves the final MAV wing type, which is the flapping wing. The flapping wing has the most potential among all of the other wing types because the flapping wing can provide both efficiency and manoeuvrability. Flapping-Wing Micro Air Vehicle, or FW-MAV is a relatively new class of MAVs that uses a wing that moves in an upward and downward cyclic motion. Since FW-MAVs are new, the number of examples of FW-MAVs is limited. However, there are plenty of examples that can be found in nature. In fact, in nature, flapping wings are the norm, while fixed wings and rotary wings are the exceptions. This is the reason flapping-wing design can also be known as bio-inspired wing design. In this study, the main focus of the MAV wing design will be the bio-inspired wing design.

One of the main issues that are involved with FW-MAV design and MAV design involves with efficiency, especially aerodynamic efficiency. Efficiency is a problem because the size restriction involved with MAVs means that the size of the power supply will also be restricted which means that the power supply will be limited. Also related, is the problem of lift generation. The limited size of the wing will also mean that the lift generation will be limited. One of the ways to allow for better lift generation is by increasing the chord length. However, increasing the chord length means increasing the drag generation. It is this balance between the lift and drag generation, and aerodynamic efficiency will be the key issue that this book aims to address. In this book, the aerodynamic efficiency is defined as the lift to drag ratio.

Bio-inspiration is a subsect of a design ideation concept called Synectic. Synectic comes from a Greek word that translated to 'to connect'. First described by Gordon (1961), Synectic can be defined as to use ideas, observation, and concepts found in a field to solve problems that are found in a different and seemingly unrelated field. Synectic can also be described as using analogy to solve a problem. Synectic can be divided into four types of analogies: direct analogy, personal analogy, fantastical analogy, and symbolic analogy. Direct analogy is the main concern for this book where it can be defined as taking ideas, observations, and concepts found in a field and directly use it in an application in a different field. One of the types of direct analogy is the concept of bio-inspiration.

Bio-inspiration is defined as taking observations found in nature and using them in a man-made problem, or in this case, in a mechanical problem. According to (Whitesides 2015), Bio-Inspiration as a concept is not particularly new. In the field of sciences and engineering, bio-inspiration can be categorised into two categories according to their motive of application. The first is biomimicry which means to copy the forms found in nature as close as possible in hopes to understand a study the phenomenon that is found in nature. The second type is bio-inspiration which

1.1 Introduction to Bio-inspired Design

means to take certain aspects found in nature and loosely copy them. This is done to find solutions for a problem that is based on the solutions that can be found in nature. While both biomimicry and bio-inspiration will continue to be discussed later in this book, the focus of the book will be mainly on bio-inspiration design.

Bio-inspiration design can also be categorised into three types. According to Benyus (2002), the three types of bio-inspirations are mimicking the forms found in nature where the geometrical shape or mechanical properties that are found in nature are used to solve a mechanical problem; the second type is mimicking the functions found in nature where concepts like the way ants communicate with each other are used to co-ordinate and control autonomous vehicles; and the third type of bio-inspiration is mimicking the environmental system found in nature like the way a forest uses waste and resources are used to design the way a city can manage its waste and resources. The main focus of bio-inspiration design approach used in this study is mimicking the forms found in nature where the shape and kinematics found in a natural wing is used to design a mechanical MAV wing design.

The main bio-inspiration design type concerns in this book is the first type of bio-inspiration design, mimicking the forms of nature. In this book, the forms found in nature refer to the physical forms that can be observed in all flying animals, mainly the wing shape (or wing morphology) and the flapping-wing kinematics. However, the sheer scale of diversity that can be found in nature means that picking a single wing type can be difficult. In nature, there are three main types of flyers: bird wings, insect wings, and batwings. Bird wings can be characterised by its skeletal feature which only exists along the span of the wing and the wing surface which is covered by the bird's feathers. In terms of wing kinematics, birds have known distance flyers since birds are found mainly outdoors and migrate to long distances. This is possible since birds are known to glide more while having a lower flapping frequency compared to other wings. The second type of wings is the insect wing type. In general, insect wings are wings that do not have a skeletal or muscles on the wing itself. Instead, the wings are driven at the root of the wing. The wing itself is just a thin membrane structure. Insect wings are known to fly in small and tight spaces since insects can be found indoors but are not known to migrate to far distances.

The wing type that is the main focus of this book is the third wing type which is the batwing type. The wing morphology and flapping kinematics will be discussed in a later chapter. However, the reason the batwing type is chosen is because bats are known to fly both indoor and over long distances. This means that the batwing type is known for both efficiency and manoeuvrability. While the bio-inspiration design can provide a solution for the design of FW-MAVs, but the approach itself still poses a few challenges.

The first and main problem that will be faced by the book involves the bio-inspiration design. One of the problems of flapping-wing design is the fact that there are few examples of a man-made flyer that utilises flapping wing. However, there are a lot of examples that can be found in nature, where flapping wings are the main form found in flying animals. One of the reasons why not more of the man-made machines are based on natural observations is because nature is overly complex. In nature, there is no single observable standard can be found where it is hard to find consistent dimensions even among the same species. Plus, simple angles and smooth surfaces do not exist in nature and the movements of wings during the flight are rarely fully consistent and are not in simple geometrical form. In a nutshell, their difficulty lies in translating observations found in nature to be applied in mechanical applications.

The second problem relates to the aerodynamics of the flapping wing itself. Flapping wings requires a complex motion during the flight that has yet to be achieved mechanically. This is because a simple sinusoidal motion will not be able to produce a stable and sustainable flight since the downward motion of the wing will produce an upward force and the upward motion of the wing will produce a downward force. This means that a simple sinusoidal flapping motion will not suffice, and a solution is needed to overcome the generated dynamic force generation of the wing. Plus, it was observed that the flapping wing is inherently high in terms of induced drag. One of the reasons why nature is successful is because the problem of the dynamic directional lift generation was overcome by using complex wing movement or complex wing deflections or complex wing shapes to maximise the favourable upward force and minimise the downward force. The problem then lies in understanding the aerodynamics of a wing with complex shape and complex dynamic motion.

The final problem relates to the design of FW-MAV itself. The problem that is inherited to MAVs lies in their limited wingspan. The limited wingspan means that the wing surface area will be limited which means that the overall lift generation is also limited. However, increasing the wing surface area will mean increasing the chord length. This then means that the drag generation will also be increased. When it comes to flapping wings, another problem will arise since flapping wings are known to exhibit increased drag induced by the flapping motion. This means that a design approach for a flapping-wing design that can generate lift while minimising the drag generation is crucial for a feasible flying FW-MAV.

The questions that this book aims to answer are:

(a) How do observations found in nature, in this case, observed found in living bats, can be translated, and used to solve mechanical problems, in this case, the problem of FW-MAV design?
(b) How do the wing geometry and wing kinematics affect the aerodynamics of a mechanical flapping wing?
(c) What are the design elements that needed to be prioritised for the design of a feasible flapping-wing micro air vehicle?

The main goal of this study is to address both problems, the aerodynamic problem associated with flapping wing, and the design problem of mimicking the complex forms found in nature.

(a) To evaluate a mechanical flapping wing that is based on observations done on natural batwings.
(b) To assess the effects of wing geometry and complaint mechanism rigidity on the wing's lift to drag ratio, leading-edge vortex, and wingtip vortex.
(c) To propose a feasible design for a batwing design with a compliant mechanism that can be used as a base for future MAV design.

Since this study focuses on bats as the source of inspiration, a restriction is needed in terms of which batwing will be used. This is because Chiroptera is an order that includes over 1000 number of species with different wing sizes and shapes. This used the wing of Cynopterus brachyotis bat as the base for the wing geometry design.

This study is still in the early development of a MAV wing. Therefore, this study will focus on a forward flight at a single fixed flight speed of 4 m/s and a steady flapping frequency of 8.5 Hz. The parameters that change are the wing geometry, wing stiffness, and angle of attack.

This study is also limited to studying the basic aerodynamic performance of the wing the measured parameters are the lift and drag. Along with the airflow visualisation that will be focused on the critical area that generates the leading-edge vortex, the trailing edge vortex, the wingtip vortex, and the wing-root vortex.

References

J.M. Benyus, *Biomimicry: Innovation Inspired by Nature* (Harper Perennial, 2002)
C. Galinski, R. Zbikowski, Some problems of micro air vehicles development. Bull. Pol. Acad. Sci.-Tech. Sci. **55**(1), 91–98 (2007)
W.J.J. Gordon, *Synectics: The Development of Creative Capacity*, 1st edn. (Harper & Brothers, 1961)
D.J. Pines, F. Bohorquez, Challenges facing future micro-air-vehicle development. J. Aircr. **43**(2), 290–305 (2006). https://doi.org/10.2514/1.4922
G.M. Whitesides, Bioinspiration: something for everyone (2015). https://doi.org/10.1098/rsfs.2015.0031

Chapter 2
Bio-inspired Flyers

Abstract This chapter is the background study where it begins with the known observations done by experts in the field of bat behaviour and morphology. The background research will be followed by a systematic study done on all the previous methodologies in the field Biomimicry and Bio-Inspired MAV wings. The background research will then end with a look at all the known aerodynamic phenomenon involved with the problem.

2.1 Introduction

According to Whitesides (2015), there are two main reasons for mechanically recreating forms found in nature. The first reason is for scientific purposes where aspects of nature like the movement of a wing are recreated as close as possible to understand the aerodynamic phenomenon that occurs during flight. This is often referred to as biomimicry. The second reason for copying nature is to use the knowledge or the observations that are found to solve existing mechanical problems. This approach is often referred to as bio-inspiration. This study will focus more on bio-inspiration by taking the observations found in bats to design a wing for MAV.

However, before any design work can be done, it is important to first understand the works that have been done and understand the current knowledge that already exists in the field. In this chapter, a background research will be laid out beginning with the current knowledge of bats and bat flight extending from the morphology of batwings to flight behaviour to flight kinematics. After that, it will be followed by a review of all the approaches in mimicking batwings and all of the wing designs inspired by bats and finding a gap in the approach that will help guide the approach that will be used in this study. This chapter will then end with a review of the aerodynamic observations that have been made in the problem of the flapping wing and bat-inspired wings and finding a knowledge gap.

© The Author(s), under exclusive license to Springer Nature Singapore Pte Ltd. 2025
S. B. Suhaimi et al., *Flapping Wing Micro Air Vehicles*,
SpringerBriefs in Applied Sciences and Technology,
https://doi.org/10.1007/978-981-96-2908-4_2

2.2 Natural Bats

The name bats, or otherwise known as *Chiroptera*, is an order of animals that consisted of 1,200 species that live all over the world. Bats are unique because they are the only mammals that are capable of sustained flight. It is due to their unique position among the animal kingdom that gives bats a unique physical feature that is not found in other animals around the world.

One of the unique features of bats is found in their wings. The name *Chiroptera* that comes from the Greek word that means 'Hand wings' that refers to the unique skeleton of bats that is like human hands have four fingers and a thumb. It is this feature that is shared among all the species of bats around the world. Other than that, the order *Chiroptera* is remarkably diverse in terms of morphology and behaviour. For example, the body range between species of order Chiroptera can range between 2 and over 1,200 g (Hubel et al. 2012). Each of the species' anatomic sizes and behaviour are determined by the environmental pressures that come from their habitats and the types of food that are available to them. In terms of diet, bats are generally known to be omnivores, but their main food source is determined by the species. Some bats are known to be feed on fruits, others are known to hunt on insects and small rodents, and there are even some species that are known to feed on nectar. These factors determine the bat's body size, wing shape, and flight behaviour.

It is this diversity that poses a problem for this study because the number of bats that is available, it is hard to choose which species can be chosen as a source of inspiration. Fortunately, the design need allows for the constraint of which species can be chosen. As mentioned before, MAVs are defined to have a wingspan of 15cm or less. However, the only species of bats that fits this criterion is Kitti's hog-nosed bat, which is known to be native to Thailand and Myanmar. Unfortunately, the Kitti's hog-nosed bat is classified as an endangered species which makes observations and studies on the bat species limited.

To overcome this problem, the scope of bat species needed to be expanded to other types of other species. One of the advantages of bio-inspiration over biomimicry is that the source of inspiration can be picked and chosen because the goal is to fulfil the design need and not to mimic a specific species of bats. In this study, instead of focusing on a single species, several works and studies that have been done have been chosen to paint an overall picture of what an average bat is capable of.

2.3 Bat Wing Morphology

As mentioned before the name *Chiroptera* comes from the Greek word 'Hand wings', bats are name *Chiroptera* because one of the main features that are shared among all of the bat species is its unique wing that resembles a human's hands. A batwing has elbow, wrist, and five digits just like a human hand. An average batwing has 25 actively controlled joints, that is at along the wingspan and the chord of the wing

2.3 Bat Wing Morphology

and have 34 degrees of freedom of motion (Bahlman et al. 2013). It is this highly morphable wing that makes batwings different from other flying animals like birds, which have actively controlled joints along the wingspan of the wing, and insects, that have actively controlled joint at the root of the wing. This gives bats manoeuvrability and flight performance that is not available for certain birds or insects.

In terms of the wing itself, four areas play a major part in lift generation which are the *Uropatagium*, the *Propatagium*, the *Plagiopatagium*, and the *Dactylopatagium*. Each of the parts plays a role in lift generation however, it is determined that during light, the centre of lift located at the *Propatagium* and the *Dactylopatagium* plays an important role in manoeuvrability.

Since this book focuses on using bats as a source of inspiration of solving a mechanical problem and not a scientific study of batwings itself, the nomenclature of shoulder, elbow, wrist, and fingers are used for simplicity. Also due to the purpose of the study to solve the mechanical design needs of a MAV, standard nomenclature for a man-made wing aircraft needed to be used and defined. According to Hubel et al. (2012), this study defines the wingspan as twice of the maximum distance between the middle of the body to the wingtip during mid-stroke or during the wing is fully stretched and flat during the flight. The chord of the wing is defined as the maximum distance between the wrist and the tip of the fifth digit (D5 in Fig. 2.1) during mid-stroke or during the wing is fully stretched and flat during the flight. The geometric wing area is defined as the area that includes the sternum (the chest of the bat), both wrists, both wingtips, both tips of the fifth digit, and the feet during mid-stroke or when the wing is fully stretched during flight.

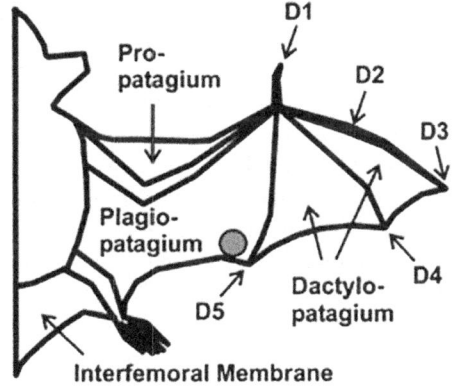

Fig. 2.1 Anatomical parts of a natural wing (Sterbing-D'Angelo et al. 2011)

2.4 Bat Flight Behaviour

As mentioned before the order *Chiroptera* is remarkably diverse and this is also true in terms of flight behaviour. For example, the species *T. Brasilliensis* are known to fly to up to 50–100 km daily for foraging and migration during autumn, while species like *G. Soricina* are known to forage within a few kilometres from their place of the roost. Bats are also diverse in terms of living behaviour with species like the *C. Brachyotis* lives in a small colony of a few hundred, while the *L. Yerbabuenae* have a large colony that has 20,000–100,000 bats (Hubel et al. 2012). While the behaviour of bats is diverse, several general patterns help to inform for MAV design. The first pattern is bats are known to live in small and confined places like caves and live in a densely populated colony. The second pattern of behaviour that is shared among bats is the fact they can forage and migrate up to 100km. This shows that bats are capable of high manoeuvrability and high endurance flight. This is especially useful for bio-inspired wing design for MAVs. Therefore, it is then useful to look at the flight behaviour of a bat.

The diverse behaviour among bats is due to several pressures. One of these factors is the type of diet, which is also diverse with some bats eat purely fruits or nectars, some hunt small rodents, and others eat insects. This study looks at a work done by Kalko (1995) on insectivore bats known as the Pipistrelle bats (or *Microchiroptera*). The reason is that insectivore bats have exhibited a high level of manoeuvrability since they need to outmanoeuvre flying insects and catch them in the air.

In the work, it was observed that insectivores forage at night and 3–11 m above ground and water surfaces along forest edges looking for small mosquito-sized insects and medium moth-sized insects. The bats normally capture their prey mid-air and the captured flight is categorised into four stages of flight: the search flight, the approach flight, the prey capture manoeuvre, and the prey retrieval.

The foraging process begins with search flight where the bat looks for its prey. During this period, the flight path is generally steady and is done in open spaces. The flight speed during this stage of the foraging process is between 4m/s and 7m/s and is found to vary depending on the size of the bats. Larger bats are found to fly faster, while smaller bats are found to fly slower.

Once prey is detected, the bat will begin to enter the approach flight and begin to pursue the prey. This phase is considered to begin when the eyes and head of the bat are re-oriented to the direction of the prey. When the bat is 30–70 cm before the prey, the bat will enter the capture manoeuvre phase where it starts to tilt its body upwards and move its tail membrane forward or extend one of its wings to scoop its prey. Once captured the bat will then bend its head into its tail pouch to retrieve the prey.

The combination of approach flight, capture flight, and prey retrieval is known as a complete pursuit manoeuvre starting from the re-orientation of the head and body and ending with the bat returning to search flight. It was observed that the entire pursuit manoeuvre lasted for 200–550 ms 97% of the time and only 3% of the time lasted for more than 1s. It was recorded that the pursuit flight speed to be between

2.4 Bat Flight Behaviour

1.5 and 3.5 m/s and at the high of the pursuit, just as the bat is about to catch the prey, the flight speed was recorded to be between 0.25 and 0.5 m/s. The flight speed varies with different types of prey and the location of the prey once it is detected, whether the prey is at the same plane of the bat's search flight, or above the bat's search flight plane, or below the bat's search flight plane. It is also interesting to note that the flight speed will reduce as it transitions from search flight to pursuit manoeuvre. This is even true when the bat dives to catch prey that is below its search flight plane. Part of the reason why is because it was recorded that the flight speed of its prey is slower than the bat with small mosquito-sized insects are found to be ranging from 0.5 to 1 m/s and medium-sized moths fly at a flight speed of between 3 and 4 m/s.

It is important to understand the flight behaviour of a bat because it helps to inform the capabilities and limitation of a batwing. The above-mentioned observation shows that bats are extremely capable of high manoeuvrability because it can dramatically change their flight speed and direction over a short period. The observed bat can change its flight speed from 4 to 1.5 to 0.25 m/s and back to 4m/s in less than 1 s. The previous observation also helps to inform the operating for future tests with 4–7 m/s for steady forward flight and 1.5–3.5 m/s for manoeuvre flight.

Another bat behaviour that shows the manoeuvrability of bats is the roosting behaviour where bats are known to roost upside down or head under heels on a surface. In a roosting manoeuvre, a bat will perform an acrobatic flip to bring claws to the ceiling while at the same time avoid injuries. According to (Riskin et al. 2009), there are two types of roosting manoeuvre: four-point landing and two-point landing. Four-point landing is defined as the roosting manoeuvre when the wrists and the feet strike the roosting surface where it begins with the pitch of the body at an average of 59.8° and then the pitch increases to 144.0° when it is nearly impacting the ceiling. Then the hind limb extends to the surface and the claws interlock to the ceiling surface. All of this goes over 0.12 s. The second type of roosting manoeuvre is the two-point landing which is defined as only the hind limbs contacted the limb where the manoeuvre begins with the pitch angle of the bat at an average of 54.7°. The pitch is then increasing, and the yaw begins a negative rotation until the feet are above the head and accompanied by a negative roll until the hind limbs reach the ceiling surface. Again, all of this goes on over the duration of 0.12 s.

It was also shown that the difference in the rate of rotation of bats during the roosting manoeuvre is dependent on the roosting surface where it was found that cave surface will make the bat to land more gently but, for foliage surface or on trees, it was found that the landing to be a lot faster. This observation is also showing that bats are capable of more manoeuvrable. Bats are capable of large changes in their pitch, roll, and yaw angle over a small amount of time.

2.5 Bat Flight Kinematics

As shown above, it was observed that bats have a flight behaviour that is very manoeuvrable while at the same time capable of flying over long distances. While ecological factors are the reason bats need to have these capabilities, the secret that makes it possible lies in the batwing motion during flight. As mentioned before another aspect that makes bats different is that their actively controlled joints are present all through the wing; both along the wingspan and the wing chord. This means that bats are capable of actively changing their wing shape far more than birds or insects. However, this also means that the wing kinematics are extremely complex, especially when the bats are achieving the previously mentioned complex manoeuvres.

For the interest of simplicity, this book will first focus on the wing kinematics of a batwing during a normal straight flight. According to Skulborstad et al. (2013), the reason for the complexity of batwing kinematics during flight is because the bats are continuously changing their shape as the flapping motion goes on. However, the wing motion can be broken down into down-stroke and upstroke motions by using the location of the wrist concerning the body. This is because, among all the wing parts, the wrist of the batwing moves the closest to a pure sinusoidal motion as possible. After all, the wrist is more directly controlled by the primary flight muscles.

During the down-stroke motion, it was observed that the wing joints are kept extended until the end of the down-stroke motion. The wings will also sweep forward as it was moving downward and this will generate the forward thrust. At the reversal point where the wing begins to transition between down-stroke and upstroke, the joints will then begin to flex, and the wingtip will move closer to the body. At the beginning of the upstroke motion, to wingtip will then move away from the body at the same time the wing moves upward. Also, during this time, the wing will sweep backwards and decrease in angle of attack as the upstroke motion is underway. The upstroke motion will end with the wing is fully extended and the cycle repeats beginning with the down-stroke motion.

Kalko (1995) recorded that the average flapping frequency of the batwing at 7.7 Hz, while Riskin et al. (2008) found that the average flapping frequency was 9.6 Hz. Much like everything else, this depends on the individual bat species and the types of flight manoeuvre are being achieved. According to Hubel et al. (2012), increasing flight speed will result in a decrease in flapping frequency. An increase in flight speed will also result in a larger difference between the upstroke wing motion and the down-stroke wind motion. At lower speeds, the flexion of the wing during the upstroke motion was found to be small. But as the flight speed increases, the wing flexes more and causes the wingtip to move closer to the body during the upstroke motion.

The complexity of the wing kinematics during flight is not only because of the large number of actively controlled joints that are involved but also because of the number of movements that happen simultaneously. Therefore, it is important to define the parameters that change during flight and the types of observable movements that are involved. According to Yu and Guan (2015), the wing motion of a bat is a combination

2.5 Bat Flight Kinematics

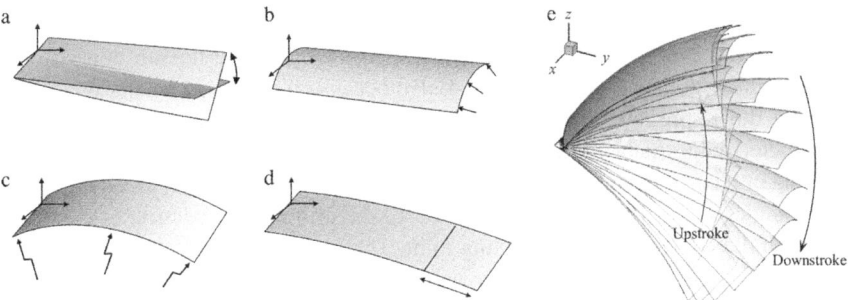

Fig. 2.2 Motions of a batwing (Yu and Guan 2015)

of several motions that occur simultaneously. Yu and Guan (2015), identified five observable motions which is twisting, cambering, bending, wing area change (or sometimes known as folding), and flapping (Fig. 2.2).

The flapping motion is the main and most dominant motion because it is the one that is most responsible for lift and thrust generation, therefore is the motion that is used as the main reference. The flapping motion is defined as the up and down movement of the wing and, if the body is used as a reference and the wingtip as a tracking point, will produce a sinusoidal pattern. Therefore, the standard parameters of measuring a sinusoidal motion are used. The wingbeat period, T, is defined as the time that it takes for a wing to complete a single flapping cycle or the time between to reversal points of the wingtip. Similarly, the flapping frequency, f, is defined as the number of flapping cycles that are completed in one second. The wing stroke amplitude, q_{tip}, is defined as the maximum angle between the locations of the wingtip to the centre body line of the bat within a given cycle. Related to the wing stroke amplitude is the flapping angle which is defined as the maximum angle between the two locations of the wingtip at the apex of down-stroke and the upstroke motions, otherwise defined as the wing stroke amplitude, q_{tip}, times two. Also, important to note, while the flapping motion moves in a sinusoidal pattern, the pattern is not purely sinusoidal. The reason is because of the other movements that were previously discussed. This results in the period for the down-stroke motion are different for the period for the upstroke motion. The difference between these periods is named the down-stroke ratio, t, which is defined as the time taken for a single down-stroke motion to complete divided by the total wingbeat period (Yu and Guan 2015).

The second motion described is the twisting motion which is defined in changing the angle of attack and have a different angle of attack along the wingspan. The geometric angle of attack of the wing, a_g, as the angle between the resultant or the flight direction and the chord line of the wing, where the chord line is defined as the line that stretches from the wrist and the trailing edge of the wing. In twisting motion, the changes are not only in terms of geometrical angle of attack but also at different parts of the wing all along the wingspan. This means that the angle of attack is also applicable to all parts of the leading edge of the wing. A third of motion is the cambering motion that is defined as the change in wing camber during the flapping flight. Wing camber is different because the change of wing camber is both active and passive. It is passive because it is changed by the flexion done by the finger digits. Camber change is also passive because it is caused by the wing skin stretches due to pressure differences generated during flight. The fourth motion that is characterised is the bending motion which is defined as the change in flex angle of the wing or when the wing bends along the wingspan. The flex angle, theta, is defined as the angle between the hand and the rest of the arm in the direction of the flight path with the wrist acts as the pivoting joint. The final motion that was described is the motion of wing area change. The wing area, in this case, is defined as the area of the wing that enclosed the sternum, wrist, wingtip, tip of the fifth digit, and the foot now of the phase-in sing stroke. This is different from the previously mentioned definition of wing surface area (geometric wing area) since this definition refers to the instantaneous change in wing surface as appose to the physical wing surface of the wing. The motion of wing area change is also known by Bahlman et al. (2014), as folding motion. The main joint that plays a huge role in wing folding is the shoulder joint and the wrist joint. The folding motion can also be defined as the change in sweep angle which is defined as the angle between the leading edge of the wing that is at the hand part of the wing and the leading edge of the wing that is at the rest of the arm part of the wing with the carpus act as a joint.

In vivo observations are important for bio-inspiration design because it works as a basis for the design work. The above previous studies serve several purposes. One of the purposes is it serves as a basis for the possible flight condition of the MAV. Conditions like possible flight speed, possible flapping frequency, and possible flapping angle. The other purpose of the previous studies is it serves as possible required wing kinematics or wing motion that is needed to be achieved for flight. However, in vivo observations also show that natural batwings are overly complex and might be beyond the possibility of the current level of technology. For example, it was shown that the wing kinematics of a batwing has an extremely high level of complexity. The change of wing geometry is unique in batwings because it has an extremely high level of active control. Even the wing skin is having an active component because according to Cheney et al. (2014), it was observed that bats have muscles in the wing skin that change the flexibility of the wing. Therefore, it is then important to review the previous attempts at recreating the forms of a batwing done by previous researchers.

2.6 Previous Studies

As mentioned before, there are two motivations for recreations of natural forms. The first motivations are for scientific purposes where machines are built as close as possible to an observed natural form to understand the physical phenomenon. The second purpose is to use the observations found in nature to solve a mechanical problem. This serves as the difference between biomimicry and bio-inspiration. However, the similarity between the two approaches lies in addressing the problem of the complexity of nature. In this study, previous works did both in biomimicry and bio-inspiration. This is because it is important to understand what is possible what is not possible for MAV design.

This study will also look and previous methods for two types of recreations, the recreation of bats wing shape, and in terms recreations of batwing kinematics. Wing flexibility is not a focus for this study because wing flexibility has the same effect of wing kinematics which is changing of wing kinematics and because as mentioned before, wing flexibility of bats is active. Therefore, wing flexibility will be considered as part of wing kinematics.

One main pattern that can be found in the approach review is how the problem of complexity is addressed. It is found that the further the work is from scientific biological study and the closer the work is to a practical MAV design, the simpler the wing design. The main pattern is addressing the complexity in nature is by simplifying the observed forms. In this review, the focus will be on how the complex forms of nature can be simplified for mechanical solutions. This book will look at the methods of simplifying the wing geometry and simplifying the flapping kinematics.

2.7 Bat Wing Geometry Mimicry

In terms of wing geometry, the complexity comes from the irregular shape of the wing itself. As mentioned before, the anatomy of a batwing is like the anatomy of a human hand. It consists of a pair of shoulders, a pair of elbows, a pair of wrists, and two sets hands with five digits fingers. The skeleton of the wing is then covered with a skin that is consists of fibrous muscles that have wrinkles. This means that the shape of a batwing is complex and highly irregular at all dimensions and simplifying the wing shape while at the same time still be true to the wing geometry of at batwing is a challenge. After studying the previous works that have done in recreating a batwing shape, it is identified that five types of recreation correlate to five levels of simplification.

The first type of batwing shape recreation is the one that is the closest to a real-life batwing shape. This method will be known by this book as the 3D realistic method of wing geometry recreation because this method aimed to recreate the wing shape as close as possible to the natural real-life wing and is based on a specific real-life bat. In this method, an in-vivo wing of a specific bat is 3D scanned and then recreated into a 3D CAD program with a certain level of realism. This method is closest to a natural batwing because this method uses a specific wing as the basis of recreation and is not based on a generalised or averaged approximation of a batwing. All the works are focused on the science of biological study of a batwing. One limitation that is found in this method is, the works that have been done are simulation studies of a batwing. Currently, there are no works that involve a fabricated wing that is based on a 3D scan of a batwing. Table 2.1 lists all works that have been done that use this method to recreate the wing.

Table 2.1 List of previous works done in 3d realistic batwing models

References	Title
Waldman et al. (2008)	Aerodynamic behaviour of compliant membranes as related to bat flight
Aono and Liu (2013)	Flapping-wing aerodynamics of a numerical biological flyer model in hovering flight
Wang et al. (2014)	Lift enhancement by dynamically changing wingspan in forward flapping flight
Joshi et al. (2020)	Full-scale aeroelastic simulations of hovering bat flight

The second type of wing geometry recreation is works that aim to recreate the skeletal structures that are observed in the wing. Once the skeletal structures are recreated, the wing skin is considered as a uniform thin membrane. This method is a slightly simplified version of the first method because the recreation effort is focused on the skeletal features of the wing. This method is also considered to be a simplification because, in this method, the wing skeletal structures are not based on any specific wing but based on an averaged and generalised observation of a batwing. Both works that have been done are test and are aimed to study the energetic costs of the flapping motion and its effects on the aerodynamic performance in a wind tunnel. All the works that have been done using this method also mainly focus on understanding the biological wing. Currently, there is still yet a flying device that uses this method to design the wing. Table 2.2 lists all works that have been done using this method (Fig. 2.3).

The third of wing recreation method is a further simplification by removing the skeletal features of the wing but keeps the general wing shape. In this method, the margin or outline shape of the wing is recreated but the wing itself is a thin wing with a constant thickness for the whole wing. However, some studies use this method but still have certain features such as wing camber, but the wing is still considered as a thin wing because the thickness of the wing is still largely uniform even when the wing is cambered. Most of the studies that use this method as a recreation method are largely focus on designing a MAV wing and studies that aim to design a MAV wing instead of done for a biological study. Wings that are designed using this method is also used for a flying MAV. This method is the highest level of wing geometry simplification where an example of a flying MAV can be found. Besides that, this method is also used in both test and simulation studies. Table 2.3 lists all works that have been done that use wings designed using this method.

The fourth method of wing shape design is a wing that removes the natural wing shape itself. In this category of wing designs, the wing shape that was used is generic or conventional wing shapes such as wings that use NACA airfoil, straight wings, elliptical wings, or thin flat wings. This book still considered wings that uses generic wings shape as a bio-inspired design because the researchers still state that the design is inspired by bats or a biological wing. Plus, some of the works that were done while uses a generic wing shape, but the works still use wing motion that is based on

2.7 Bat Wing Geometry Mimicry

Fig. 2.3 Wing models with skeletal features and membrane wings (Bahlman et al. 2013)

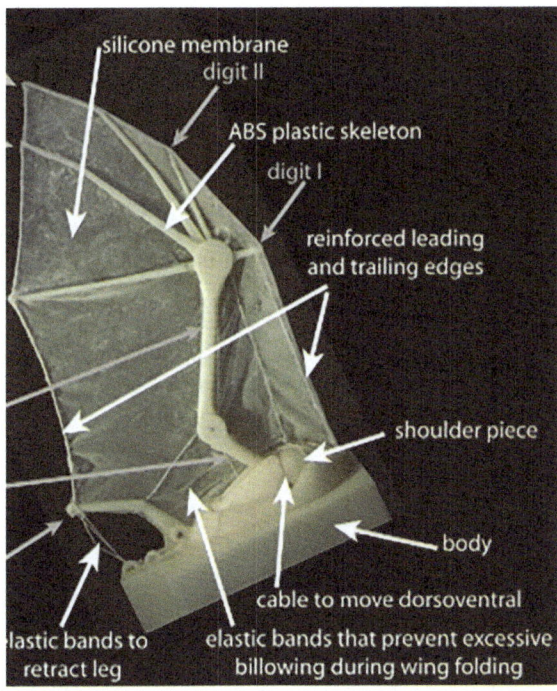

Table 2.2 List of previous works done in wing models with skeletal features

References	Title
Colorado et al. (2012)	Biomechanics of smart wings in a bat robot : morphing wings using SMA actuators
Bahlman et al. (2013)	Design and characterization of a multi-articulated robotic batwing
Bahlman et al. (2014)	How wing kinematics affect power requirements and aerodynamic force production in a robotic batwing
Stowers and Lentink (2015)	Folding in and out: passive morphing in flapping wings
Yin and Zhang (2016)	Learning from bat: aerodynamics of actively morphing wing
Zhang et al. (2018)	Design of a hydraulically-driven bionic folding wing
Duan et al. (2023)	Wing geometry and kinematic parameters optimization of bat-like robot fixed-altitude flight for minimum energy
Bouard et al. (2024)	Aerodynamics of flapping wings with passive and active deformation

batwings. Generic wings were used for biomimicry studies for studying the effects of certain adjustable factors or to focus on certain aerodynamic phenomena. By using a conventional wing, the researchers were able to remove other external adjustable factors such as the effect of the natural wing geometry. Works that use wing designs that fall into this category of design are simulation studies, test works, and flying

Table 2.3 List of previous works done with the margin shape model

References	Title
Pornsin-Sirirak et al. (2000)	MEMS wing technology for a battery-powered ornithopter
Pornsin-Sirirak et al. (2001)	Titanium-alloy MEMS wing technology for a micro aerial vehicle application
Pornsin-Sirirak et al. (2001)	Microbat: a palm-sized electrically powered ornithopter
Yusoff et al. (2015)	Lift performance of a cambered wing for aerodynamic performance enhancement of the flapping wing
Yusoff et al. (2015)	Test study on the effect of skin flexibility on aerodynamic performance of flexible skin flapping wings for micro air vehicles
Cong et al. (2023)	Aerodynamic performance of low aspect-ratio flapping wing with active wing-chord adjustment
Liu et al. (2024)	Effects of dynamical spanwise retraction and stretch on flapping-forward flights
Zhang et al. (2024)	Analysis of the integrated pattern of hoverable flapping wing micro-air vehicle

devices. Table 2.4 lists all the works did that uses wing designs that fall into this category (Fig. 2.4).

The final method or category is works that use 2D wings were the wings that are either NACA aerofoils or thin wings with features like wing cambering. This

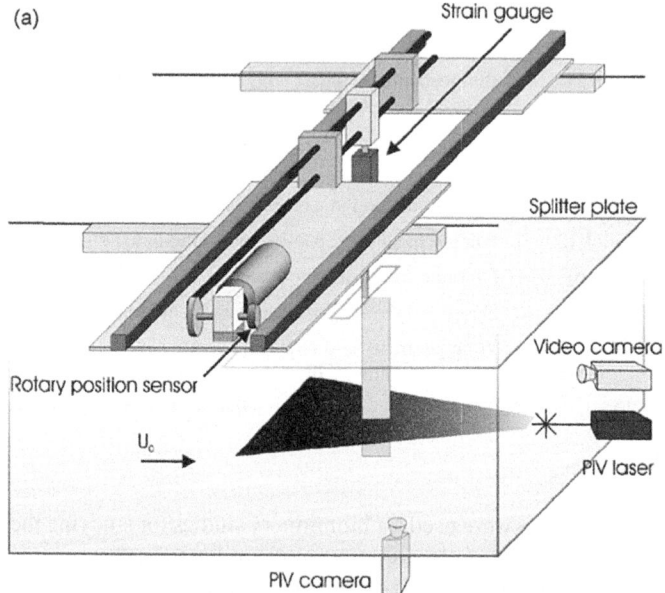

Fig. 2.4 Wing models that use conventional or generic wing shapes (Heathcote et al. 2008)

2.7 Bat Wing Geometry Mimicry

Table 2.4 List of previous works done with conventional or generic wing shapes

References	Title
Heathcote et al. (2008)	Effect of spanwise flexibility on flapping wing propulsion
Le et al. (2010)	Effect of chord flexure on aerodynamic performance of a flapping wing
De Rosis (2014)	On the dynamics of a tandem of asynchronous flapping wings: lattice Boltzmann-immersed boundary simulations
Fenercioglu and Cetiner (2014)	Effect of unequal flapping frequencies on flow structures
De Rosis et al. (2014)	Aeroelastic study of flexible flapping wings by a coupled lattice Boltzmann-finite element approach with immersed boundary method
Deng et al. (2014)	Test investigation on the aerodynamics of a bio-inspired flexible flapping wing micro air vehicle
Wenqing et al. (2015)	Aerodynamic performance of micro flexible flapping wing by numerical simulation
Cheng and Lan (2015)	Effects of chordwise flexibility on the aerodynamic performance of a 3D flapping wing
Hoke et al. (2015)	Effects of time-varying camber deformation on flapping foil propulsion and power extraction
Tay (2016)	Effect of different types of wing-wing interactions in flapping Mavs
Bleischwitz et al. (2016)	Aeromechanics of membrane and rigid wings in and out of ground-effect at moderate Reynolds numbers
Moriche et al. (2016)	Three-dimensional instabilities in the wake of a flapping wing at low Reynolds number
Sun and Zhang (2017)	Effect of the reinforced leading or trailing edge on the aerodynamic performance of a perimeter-reinforced membrane wing
Bleischwitz et al. (2017)	On the fluid-structure interaction of flexible membrane wings for Mavs in and out of ground-effect
Huera-Huarte and Gharib (2017)	On the effects of tip deflection in flapping propulsion
Wu et al. (2017)	Automated kinematics measurement and aerodynamics of a bioinspired flapping rotary wing
Guo et al. (2018)	Analysis and test of a bio-inspired flyable micro flapping wing rotor
He et al. (2018)	Modeling and vibration control of the flapping-wing robotic aircraft with output constraint
Wen et al. (2018)	Nonlinear dynamics of a flapping rotary wing: modeling and optimal wing kinematic analysis
Chen et al. (2019)	Structural integrity analysis of transmission structure in flapping-wing micro aerial vehicle via 3d printing
Gong et al. (2019)	Numerical investigation of the effects of different parameters on the thrust performance of three dimensional flapping wings

(continued)

Table 2.4 (continued)

References	Title
Dong et al. (2020)	Design and test study of a new flapping wing rotor micro aerial vehicle
Isakhani et al. (2020)	Fabrication and mechanical analysis of bioinspired gliding-optimized wing prototypes for micro aerial vehicles
Bie et al. (2021)	Design, aerodynamic analysis and test flight of a bat-inspired tailless flapping wing unmanned aerial vehicle
Lahoti et al. (2022)	Design and development of a folding mechanism for bat-like bioinspired wing
Kan et al. (2023)	Design and flight test of the fixed-flapping hybrid morphing wing aerial vehicle
Pfliger and Breitsamter (2024)	Gust response of an elasto-flexible morphing wing using fluid-structure interaction simulations
Torregrsosa et al. (2024)	Multifidelity approach to the numerical aeroelastic simulation of flexible membrane wings

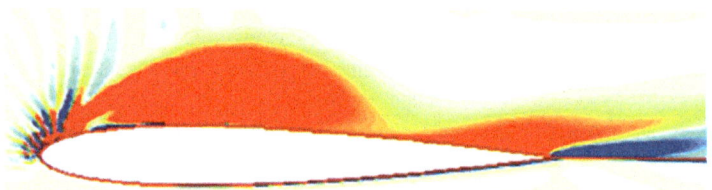

Fig. 2.5 2D wing models (Geissler and van der Wall 2017)

study includes this type of wings because the studies that use these wings focuses on an aerodynamic phenomenon that is relevant to the design of a MAV wing, or the works cited batwings as its source of design inspirations, or the focus is on recreating the wing motion of a batwing. Due to the nature of the wing designs, it is impossible for test work or a flying device. Works that use this type of wings are exclusively simulation studies. Table 2.5 lists all works did that uses wings that fall in this category (Fig. 2.5).

2.8 Wing Kinematics Mimicry

As mentioned before, the wing motions or kinematics of bats during the flight are overly complex due to the number of motions and changing angles that all occur simultaneously and at different speeds. The complex motions mean that recreating a one to one movement is exceedingly difficult. Similar to wing geometry recreation, simplifications was done for the wing motion be achievable. This book identified five categories or levels of simplification of wing kinematic recreation.

2.8 Wing Kinematics Mimicry

Table 2.5 List of previous works done in 2D wing models

References	Title
Wu et al. (2015a, b)	Ground effect on the power extraction performance of a flapping wing biomimetic energy generator
Wu et al. (2015a, b)	How a flexible tail improves the power extraction efficiency of a semi-activated flapping foil system: a numerical study
Du and Sun (2015)	Effect of flapping frequency on aerodynamics of wing in freely hovering flight
Olivier and Dumas (2016a)	A parametric investigation of the propulsion of 2D chordwise-flexible flapping wings at low Reynolds number using numerical simulations
Olivier and Dumas (2016b)	Effects of mass and chordwise flexibility on 2D self-propelled flapping wings
Lua et al. (2016)	On the thrust performance of a flapping two-dimensional elliptic airfoil in a forward flight
Chen et al. (2017)	A fully-activated flapping foil in wind gust: energy harvesting performance investigation
Geissler and van der Wall (2017)	Dynamic stall control on flapping wing airfoils
Rahman and Tafti (2020)	The role of vortex–vortex interactions in thrust production for a plunging flat plate

The first type of wing kinematic recreation is the most realistic where motion tracking of a specific individual batwing was done in a laboratory environment and the tracking data was then used to recreate the wing motion in a simulation study. This method is used largely to study the aerodynamics of in vivo bat flight in a CFD simulation in place of an in vivo wind tunnel testing. However, there is no example of this method used for a mechanical or engineering application purposes. Also, there is not yet an example of this method is recreated in an test capacity due to the complex nature of the wing motion. Figure 2.6 shows an example of motion recreation using motion tracking data where illuminated dots located at certain parts of the bat where the dots were used to track the movement of the bats during flight and then the motion data transferred for a CFD study tracking (Tian et al. 2006). Table 2.6 lists all studies that have been done that use this method.

The second type of wing recreation is a simplified and more generalised version of the first wing kinematics recreation method where instead of the wing motion is based on an individual batwing kinematic, the recreation is based on a generalised observation of bat flight. This is done by trying to achieve a combination of all more than two motions that are observed in a bat flight. Instead of using tracking data for wing kinematic recreation, this method attempts to recreate the wing bending, wing folding, wing cambering, wing twisting, along with the flapping motion of the wing. This method is the highest level of simplification where an test study can be found along with a simulation study. However, there is not yet an example of a flying device that uses this method for wing kinematics. This is because the wing motion is still complex enough where it is still too difficult to recreate the wing

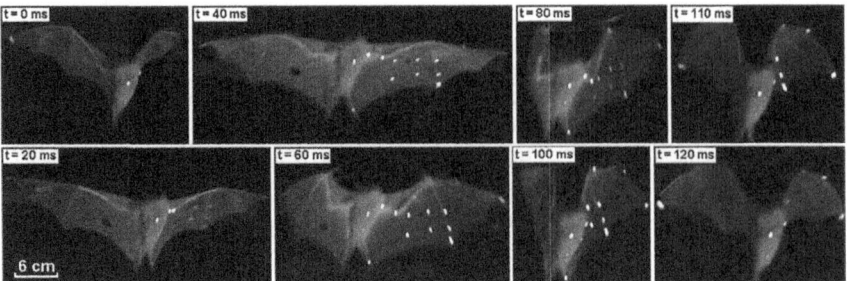

Fig. 2.6 Wing motion model based on natural wing motion tracking (Tian et al. 2006)

Table 2.6 List of previous works done with natural wing motion tracking

References	Title
Tian et al. (2006)	Direct measurements of the kinematics and dynamics of bat flight
Riskin et al. (2008)	Quantifying the complexity of batwing kinematics
Wang et al. (2014)	Lift enhancement by dynamically changing wingspan in forward flapping flight
Meng and Sun (2016)	Wing kinematics, aerodynamic forces and vortex-wake structures in fruit-flies in forward flight
Joshi et al. (2020)	Full-scale aeroelastic simulations of hovering bat flight

motion where it moves fast enough or to create a device that light enough where the flight can be achieved. Because of that also this method is only used for biological study purposes and not for any engineering application. Figure 2.7 shows an example of wing mechanisms that uses this motion recreation where a series of pulleys and motors used to recreate a the flapping motion combined with folding and bending motion (Bahlman et al. 2014). Table 2.7 lists all the works that use this method of wing kinematics.

The third type of wing kinematics recreation is an even more simplified version of the second method where, instead of recreating a two or more combination of wing motion along with the flapping motion, the third type of wing kinematic recreation only combines one type of motion with the flapping motion. The type of motion can be wing bending, cambering, twisting, or folding, but only one of them combined with the flapping motion. This method can be advantageous over the previous method because it helps to highlight the contributions of each of the wing motion to the flapping flight. This method is also the first level where a flying device can be found. This method is also used in test and simulation studies, and this method is also used in studies for both biological and engineering works. Table 2.8 lists all the works that use this type of wing kinematics.

The fourth type of wing kinematics is an even further level of simplification where instead of a combination of other wing motion with flapping motion, the fourth type of wing motion focuses wing-flapping motion entirely. The pure flapping motion is

2.8 Wing Kinematics Mimicry

Fig. 2.7 Wing motion model that combines several wing motion types (Bahlman et al. 2014)

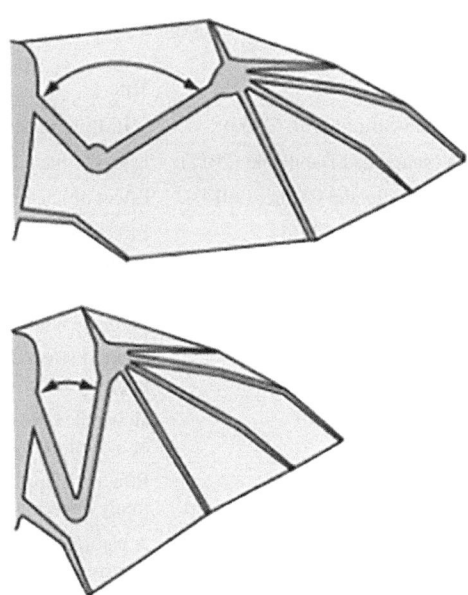

Table 2.7 List of previous works done with flapping motion combined with several motion types

References	Title
Bahlman et al. (2013)	Design and characterization of a multi-articulated robotic batwing
Aono and Liu (2013)	Flapping wing aerodynamics of a numerical biological flyer model in hovering flight
Bahlman et al. (2014)	How wing kinematics affect power requirements and aerodynamic force production in a robotic batwing
Yin and Zhang (2016)	The inertial power and inertial force of robotic and natural batwing
Li et al. (2016)	Unsteady aerodynamic and optimal kinematic analysis of a micro flapping wing rotor
Zhang et al. (2018)	Design of a hydraulically-driven bionic folding wing
Liu et al. (2024)	Effects of dynamical spanwise retraction and stretch on flapping-forward flights

considered in this book because at its purest form a bat flight is flapping flight and studies that use flapping motion helps to understand the nature of bat flight in its simplest form by studying the effects of changing flapping frequency and changing the flapping angle. This type of study is possible for all flying, test, and simulation studies and can be used for biological and engineering purposes. Figure 2.8 shows an example of work that uses his method where a servo shaft was used to create a pure flapping motion for an test study. Table 2.9 lists all of the previous works that have done that use this model approach.

Table 2.8 List of previous works done with flapping motion combined with one other wing motion types

References	Title
Pornsin-Sirirak et al. (2000)	MEMS wing technology for a battery-powered ornithopter
La Mantia and Dabnichki (2011)	Effect of the wing shape on the thrust of flapping wing
Fenercioglu and Cetiner (2014)	Effect of unequal flapping frequencies on flow structures
Cheng and Lan (2015)	Effects of chordwise flexibility on the aerodynamic performance of a 3D flapping wing
Hoke et al. (2015)	Effects of time-varying camber deformation on flapping foil propulsion and power extraction
Wu et al. (2015a, b)	Ground effect on the power extraction performance of a flapping wing biomimetic energy generator
Wu et al. (2015a, b)	How a flexible tail improves the power extraction efficiency of a semi-activated flapping foil system: a numerical study
Du and Sun (2015)	Effect of flapping frequency on aerodynamics of wing in freely hovering flight
Olivier and Dumas (2016a)	A parametric investigation of the propulsion of 2D chordwise-flexible flapping wings at low reynolds number using numerical simulations
Olivier and Dumas (2016b)	Effects of mass and chordwise flexibility on 2D self-propelled flapping wings
Moriche et al. (2016)	Three-dimensional instabilities in the wake of a flapping wing at low Reynolds number
Lua et al. (2016)	On the thrust performance of a flapping two-dimensional elliptic airfoil in a forward flight
Huera-Huarte and Gharib (2017)	On the effects of tip deflection in flapping propulsion
Chen et al. (2017)	A fully-activated flapping foil in wind gust: energy harvesting performance investigation
Geissler and van der Wall (2017)	Dynamic stall control on flapping wing airfoils
Wu et al. (2017)	Automated kinematics measurement and aerodynamics of a bioinspired flapping rotary wing
Guo et al. (2018)	Analysis and test of a bio-inspired flyable micro flapping wing rotor
He et al. (2018)	Modelling and vibration control of the flapping-wing robotic aircraft with output constraint
Jankauski et al. (2018)	The effect of structural deformation on flapping wing energetics
Wen et al. (2018)	Nonlinear dynamics of a flapping rotary wing: modeling and optimal wing kinematic analysis
Chen et al. (2019)	Structural integrity analysis of transmission structure in flapping-wing micro aerial vehicle via 3D printing
Gong et al. (2019)	Numerical investigation of the effects of different parameters on the thrust performance of three dimensional flapping wings

(continued)

2.8 Wing Kinematics Mimicry

Table 2.8 (continued)

References	Title
Dong et al. (2020)	Design and test study of a new flapping wing rotor micro aerial vehicle
Rahman and Tafti (2020)	The role of vortex–vortex interactions in thrust production for a plunging flat plate
Lahoti et al. (2022)	Design and development of a folding mechanism for bat-like bioinspired wing
Duan et al. (2023)	Wing geometry and kinematic parameters optimization of bat-like robot fixed-altitude flight for minimum energy
Zhang et al. (2024)	Analysis of the integrated pattern of hoverable flapping wing micro-air vehicle

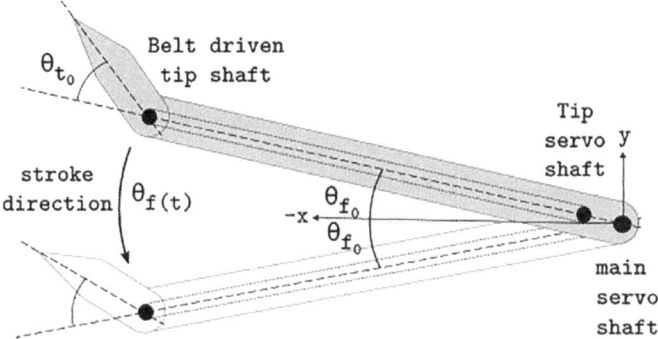

Fig. 2.8 Wing motion model that uses pure flapping motion (Huera-Huarte and Gharib 2017)

The fifth and final type of wing kinematic is the fixed-wing motion. Studies that use fixed-wing is considered in this book because while the studies omitted all wing motions including the flapping motion, the works done studied the effects of wing geometry and the effects of wing skin which can be useful. Plus, the gliding flight is still a possibility for a bat although it is not as common as flapping flight. However, this book does not qualify flying devices that use fixed-wing because it is beyond the scope of the book. Therefore, the only works did that is considered in this book that uses fixed-wing are works that are both test and simulation done for biological and engineering purposes. Figure 2.9 shows an example of a work done for fixed-wing studies where a CFD simulation was done for a fixed-wing at a set wind speed (Wang et al. 2017). Table 2.10 lists all the previous works did that uses fixed-wing.

Table 2.9 List of previous works done with a pure flapping motion

References	Title
Pornsin-Sirirak et al. (2001)	Titanium-alloy MEMS wing technology for a micro aerial vehicle application
Pornsin-Sirirak et al. (2001)	Microbat: a palm-sized electrically powered ornithopter
Waldman et al. (2008)	Aerodynamic behavior of compliant membranes as related to bat flight
Heathcote et al. (2008)	Effect of spanwise flexibility on flapping wing propulsion
Le et al. (2010)	Effect of chord flexure on aerodynamic performance of a flapping wing
Colorado et al. (2012)	Biomechanics of smart wings in a bat robot : morphing wings using SMA actuators
Tian et al. (2013)	Force production and asymmetric deformation of a flexible flapping wing in forward flight
Altshuler et al. (2004)	Aerodynamic forces of revolving hummingbird wings and wing models
De Rosis et al. (2014)	Aeroelastic study of flexible flapping wings by a coupled lattice Boltzmann-finite element approach with immersed boundary method
Deng et al. (2014)	Test investigation on the aerodynamics of a bio-inspired flexible flapping wing micro air vehicle
Yusoff et al. (2014)	Test study on the effect of skin flexibility on aerodynamic performance of flapping wings for micro air vehicles
Yusoff et al. (2015)	Test study on the effect of skin flexibility on aerodynamic performance of flexible skin flapping wings for micro air vehicles
Wenqing et al. (2015)	Aerodynamic performance of micro flexible flapping wing by numerical simulation
Cleaver et al. (2016)	Lift enhancement through flexibility of plunging wings at low Reynolds numbers
Tay (2016)	Symmetrical and non-symmetrical 3D wing deformation of flapping micro aerial vehicles
Huera-Huarte and Gharib (2017)	On the effects of tip deflection in flapping propulsion
Li et al. (2018)	A novel 3D variational aeroelastic framework for flexible multibody dynamics: application to bat-like flapping dynamics
Lahoti et al. (2022)	Design and development of a folding mechanism for bat-like bioinspired wing
Cong et al. (2023)	Aerodynamic performance of low aspect-ratio flapping wing with active wing-chord adjustment
(2023)	Design and flight test of the fixed-flapping hybrid morphing wing aerial vehicle

(continued)

2.8 Wing Kinematics Mimicry

Table 2.9 (continued)

References	Title
Bouard et al. (2024)	Aerodynamics of flapping wings with passive and active deformation
Torregrsosa et al. (2024)	Multi-fidelity approach to the numerical aeroelastic simulation of flexible membrane wings

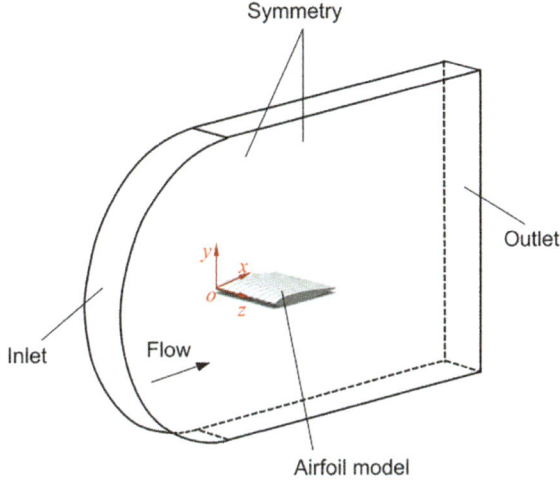

Fig. 2.9 Fixed-wing motion model (Wang et al. 2017)

Table 2.10 List of previous works done with fixed-wing

References	Title
Bleischwitz et al. (2016)	Aeromechanics of membrane and rigid wings in and out of ground-effect at moderate reynolds numbers
Sun and Zhang (2017)	Effect of the reinforced leading or trailing edge on the aerodynamic PERFORMANCE OF A PERIMETER-REINFORCED MEMBRANE WING
Bleischwitz et al. (2017)	On the fluid-structure interaction of flexible membrane wings for mavs in and out of ground-effect
Wang et al. (2017)	Numerical study on reduction of aerodynamic noise around an airfoil with biomimetic structures
Isakhani et al. (2020)	Fabrication and mechanical analysis of bioinspired gliding-optimized wing prototypes for micro aerial vehicles
Pfliger and Breitsamter (2024)	Gust response of an elasto-flexible morphing wing using fluid-structure interaction simulations

2.9 Wing Aerodynamics

The second part of this book focuses on the aerodynamic problem that is involved with flapping-wing MAV design. Therefore, what is needed is a review of the current knowledge of the aerodynamic phenomenon that occurs in a flight of a wing design based on bats. The review begins with an analysis of the observed flow pattern generated by an in-vivo bat flight. This is then followed by the effects of several design factors on the aerodynamics of the wing. Finally, the review will end on the flow pattern that is key in aerodynamic performance and how it is achieved.

Since this book focuses heavily on bio-inspiration design, it important to first look at the observed aerodynamic flow pattern of an in-vivo bat flight to inform the later mechanical designs. According to Hubel et al. (2010) the main generator of lift for a bat flight comes from the down-stroke motion and the bat flight itself begins with the down-stroke motion. It is observed that the keys to lift generation are the WTV and the LEV. In terms of WTV, the vortex generated during the majority of the down-stroke motion and continues into the upstroke motion. The vortex begins to appear at the first third of the down-stroke motion and the vortex begin to gain in strength as the sown-stroke motion goes on. The vortex will also begin to detach from the wing and begins to move downward and outward. The vortex was found to be at its peak strength at the end of the down-stroke and the beginning of the upstroke. At the beginning of the upstroke wing motion, the vortex was found to upward and into the wing. The vortex remains during the upstroke motion until the halfway point of the wing stroke motion where the vortex will begin to weaken and die out until no circulation can be found at the end of the upstroke motion. Also observed during the upstroke motion, is a short period where a reverse circulation is observed at the wingtip that generates a negative lift and explains the decay of the vortex strength (Hubel et al. 2012).

In terms of the LEV, however, it was observed that it is key during low-speed flights. However, it is important to note that the aerodynamic performance is also dependant on the flight speed. Hubel et al. (2012) has shown that at low flight speeds, the induced drag is observed to be lower and this means that the lift is more dependent on the wing surface area with higher wing area generates higher lift. At higher flight speeds, however, it was shown that profile drag, and parasitic drag was observed to be more important and this means that an increase in wing area has a detrimental effect on the aerodynamic performance. This also explains why the wing extends less at flight speed.

While in-vivo observation helps in informing the aerodynamics of a bat flight there are still several limitations to pure in-vivo observations. One of which is the fact that the results can vary between individual species because there are no two bats that have the same body mass and body build. The second limitation to in-vivo observation is it is difficult to produce a humane and ethical approach that tests the different factors of flight performance. The only observations will be limited to the natural flight pattern of a bat at its natural forms. Therefore, this book will have to rely on studies that are done in an test and simulation setting.

2.9 Wing Aerodynamics

Another aspect of MAV design that needed to be covered by this book is the effect of flight conditions on the aerodynamic performance of a MAV flapping wing. As mentioned before. In-vivo studies on bats have found that the flapping kinematics of a batwing will change with flight speed. As the flight speed increases, the wing beat amplitude and the vertical acceleration increases while the flapping frequency, sweep angle, flex angle, down-stroke ratio, wingspan, and average angle of attack decreases (Hubel et al. 2012).

The reason for the change in wing kinematics is because flight conditions like flapping frequency and flight speed influence the aerodynamics performance. In terms of flight speed, according to De Rosis (2014), at a fixed flapping frequency, increasing flight speed will cause a detrimental effect on the aerodynamic performance. However, if the flapping frequency increases at fixed flight speed, it was found that the thrust will increase (Deng et al. 2014) and the drag will decrease (Wu et al. 2014). The high lift can be achieved with a low and high frequency, but the lift will be decreased at a certain ill-fit medium frequency. The reason for this strange relationship between flapping frequency, flight speed and the aerodynamic performance is because the aerodynamic performance is affected by a ratio of flight speed and flapping frequency called reduced frequency. As the reduce frequency increases, the unsteady effects of the wing increases. With decreasing forward flight speed and the angular velocity increases, the unsteady effects also increase which means that the role of vortices in lift generation and stall angle delay increases (Shyy et al. 2008).

According to Yusoff et al. (2014), it was found that with increasing advance ratio the lift will decrease, and the drag will increase until at a certain point where the lift and drag remain constant with changing advance ratio. This relationship between the advance ratio and aerodynamic performance can be explained by an in-vivo study done by Muijres et al. (2012) where it was found that bats have better performance than birds in a flapping flight at lower speeds but birds are more efficient in a glide flight at higher speeds. This is the reason for bats are rarely found to do glide flight since a batwing is more efficient in a flapping dominant low flight speed.

In terms of the effect of the wing shape on the aerodynamics performance, some of the usual conventions found in conventional wings are found to still stand for a bio-inspired flapping wing. For instance, according to Yu and Guan (2015), it was found that with increasing wing surface area the lift of the wing is also increased and with decreasing wing area, the generated lift is also decreased. It is also was observed that the thickness of the wing can also affect the wing aerodynamic performance where with increasing wing thickness the drag will also increase and the lift will decrease (Xiang et al. 2016). However, flapping-wing differs from a conventional wing because one of the unique features of a bio-inspired flapping-wing is the fact that the wing itself is constantly in motion and unlike rotary wings, most flapping-wing undergoes a certain type of wing shape changes which can come in two forms; passive morphing and active morphing.

One of the problems in bio-inspired design that this book aims to address is the problem of translating from the biological observation to the mechanical application. One of the ways that this problem arises is in the problem of active and passive wing shape morphing. The reason this is a problem is that there is no clear answer to whether or not the wing shape changes is due to external forces or due to an active change made by the bat itself. Wing skin morphing is often considered as a passive wing morphing since this is true for birds and insects. But bats are different because of the presence of muscles in the skin that allows the bat to change its flexibility during flight. Wing joints movements are often considered active wing shape morphing since that is the main purpose of the arm muscles in the batwing. But bats are also known for passive flexure during flight by relaxing the muscles and letting the air pressure to bends and fold during flight. When it comes to translating the mechanical problem, however, simplification has made it easier since, aside from several cases, it is considered passive and it is considered active has been made a lot clearer. This is important because it justifies looking at works that focus on subjects that is other than the focus of this book. For instance, this book will also focus on works that are focused on the wing that is other than batwings. This is because there are limitations of knowledge especially in terms of aerodynamics phenomenon but there are works done in other types of wings that explore these areas. The other studies on other wings valid for this book because there are similarities that are there and can help to inform the work that is being done. This also means this book will review works that focus on other factors that are not immediately obvious to be the focus of this book (factors like, wing flexibility) but is still being reviewed because the information is important to paint a better picture of understanding the aerodynamics of batwing flight.

In terms of the effect of wing stability on the aerodynamic performance, several studies (Song et al. 2008; Yusoff et al. 2015) have shown that the more flexible a wing will produce better lift and higher stall angle. This is because according to (Aono and Liu 2013), flexible wings can generate and stabilize the LEV which, as mentioned before plays a role in lift generation. This observation is consistent with other works that use generic wings (Cleaver et al. 2016; De Rosis 2014; Wu et al. 2015a, b) but works done by Cheng and Lan (2015), Yang et al. (2018) seems to contradict this observation by observing that the increasing wing flexibility can cause lift degradation. This contradiction is then expanded by Heathcote et al. (2008) where it was observed that there is a range of effective wing flexibility for lift generation. However, pure global flexibility does not reveal the whole picture and local flexibility and wing flexure also have a huge effect on on-wing performance. According to Jeanmonod and Olivier (2017), Tay et al. (2018), it is observed that span-wise flexibility and chord-wise flexibility influence the thrust generation of the wing. It was also observed that front flexibility is better in a pressure-driven situation while rear flexibility is better for inertia driven situations. Optimum wing flexure has been studied by Lee et al. (2016), where it was observed that thrust and lift of a wing are dependent on how the wing flexure amplitude.

The reason wing flexibility affects the lift, drag, thrust, and efficiency of the wing comes from the ability of the wing to generate Leading Edge Vortex (LEV). According to Sun and Zhang (2017), the LEV is generated especially at lower angles of attack by the interaction between the leading edge of the wing and the forward airflow. As the angle of attack increases, it was found that another weaker vortex forms and pushes the LEV away from the wing surface causing the lift to decrease.

2.9 Wing Aerodynamics

Hoke et al. (2015), explains that the advantage of flexible wings is its ability to morph the wing skin with the changing aerodynamic pressure and keeping the LEV to stay on the wing surface for a lot longer. This explains why flexible wings can have better aerodynamic efficiency and have higher stall angles and why front flexible wings are found to be better for pressure-driven situations. This also explains why in in-vivo bats, the LEV is more present in slower flight speeds where the wing is found to be more pressure driven. On the other hand, (Heathcote et al. 2008), have found that too much flexibility can be detrimental to the lift generation. This is because if a wing is too flexible, this will cause the wing to generate weak vorticity, especially the leading-edge vorticity. Therefore, it is also important to keep part of its shape and integrity for lift generation to be possible.

Another main aspect of lift generation lies in the kinematics of the flapping bat itself. As mentioned before, the kinematics of a wing motion is overly complex with several wing four types of wing motion (twisting, cambering, bending, and folding) that occurs at the same time as the flapping motion. It was also mentioned that the wing movements are different during the down-stoke period then the movements during the upstroke period. Plus, each of the wing motions plays a role in aerodynamic performance. The twisting motion, for example, it was found that the twisting motion plays a role in increasing the overall time-averaged lift by decreasing the difference in lift generation of the down-stroke and upstroke motion. The twisting motion was also observed to help with thrust generation. The twisting motion enhances the thrust and lifts by modulating the vortex shedding at later periods of the down-stroke motions and early periods of the upstroke motion by increasing the amplitude of the trailing edge.

In terms of wing cambering, however, it was observed that wing cambering can increase the lift and thrust, especially during the down-stroke period. Although it was also observed that wing camber can also cause the drag of the wing to increase especially if the wing camber exists during the upstroke period (Yusoff et al. 2015). This is because the wing camber increases the effective angle of attack of the wing during the down-stroke motion which results in increasing lift generation. It is also due to the wing cambering allows for the generated vortex to leave the wing surface later in the down-stroke process.

The wing motions that arguably have the biggest role for a bat flight are the wing bending and the wing folding motion. Both motions are similar because they both essentially play the same role, changing the effective area of the wing. These two motions are arguably considered to play the biggest role because these two motions are the ones that change the wing shape the most. According to Waldman et al. (2008), it was observed that the difference between the effective wing area during down-stroke motion and the upstroke motion is estimated to be 100%. This means that that the wing area during the down-stroke period is twice as much as the wing area during the upstroke period. The changing of the wing area plays a vital role in minimising the negative or adverse effect of the upstroke motion on the generated lift and drag. The larger the wing surface area, the more lift generated during the down-stroke and the more adverse effect it has during the upstroke. By making the wing surface area smaller during the upstroke motion, the negative lift, or the adverse effect the upstroke motion has on the lift and drag.

Again, upon reviewing the factors that affect the aerodynamic performance of a batwing, it has shown that there are blurred lines the active and passive change in wing geometry. While the wing kinematics is generally considered to be an active motion, some of its effects can be achieved with passive methods and can produce the same results. For instance, the wing twisting motion can be achieved by actuating the wing at the shoulder, but the same thing can be achieved by having a wing that is flexible along the wingspan. The wing change in wing camber can be achieved by wither flexing one of the digits in the wing or by having the wing to be flexible in the chord-wise direction. The wing bending can be achieved by actuating the wing at the elbow of the wing or just by letting the wing to be flexible enough span-wise, that it allows for wing flexure. The only motion that is not possible for a passive solution is folding. Both methods either passive or active can achieve the same wing morphing and wing aerodynamics results.

It is at this point where it is useful to focus on the main objective of the study which is to produce a wing design of a MAV. One of the main hindrances of bio-inspired flying device design is complexity. Active wing motion is one of the main sources of complexity for a MAV design. This is because an active motion will require parts for the actuation of the active motion. Therefore, one of the keys for simplification in MAV design is keeping the number of required active motion actuation as possible and allowing for as much passive motion as possible, and not focusing on factors that does not have as much of an impact.

2.10 Knowledge Gap

For the review of the previous works to serve the purpose of this book, it is first useful to look at the main objectives and focus of the book first. The focus of the book can be divided into two parts. The first part is the design question, which is how an observation from a natural wing-like batwings does can be used to solve a mechanical problem like the design of a flapping-wing MAV. The second part is the aerodynamic question, which is what are the main aerodynamic phenomenon and factors that are key in designing a flapping wing for a flapping-wing MAV (Fig. 2.10).

For the design question, the problem stems from the complexity of a batwing in terms of wing shape and in terms of kinematics that makes a one-to-one mimicry impractical for the creation of a flying MAV. The approach that is key to solving this problem is by simplifying the observations found in an in-vivo study. In the table above, a matrix of previous works from the year 2000 to 2020 that studies different types of flapping wings which includes bat wings, bird wings, insect wings, and generic wings.

Upon reviewing previous works done, it is found that a flat wing with a margin shape of a batwing is the wing type that is the closest to an in-vivo wing but still able to produce a flying device. However, the wing margin wing shape can also

2.10 Knowledge Gap

Fig. 2.10 Previous study matrix

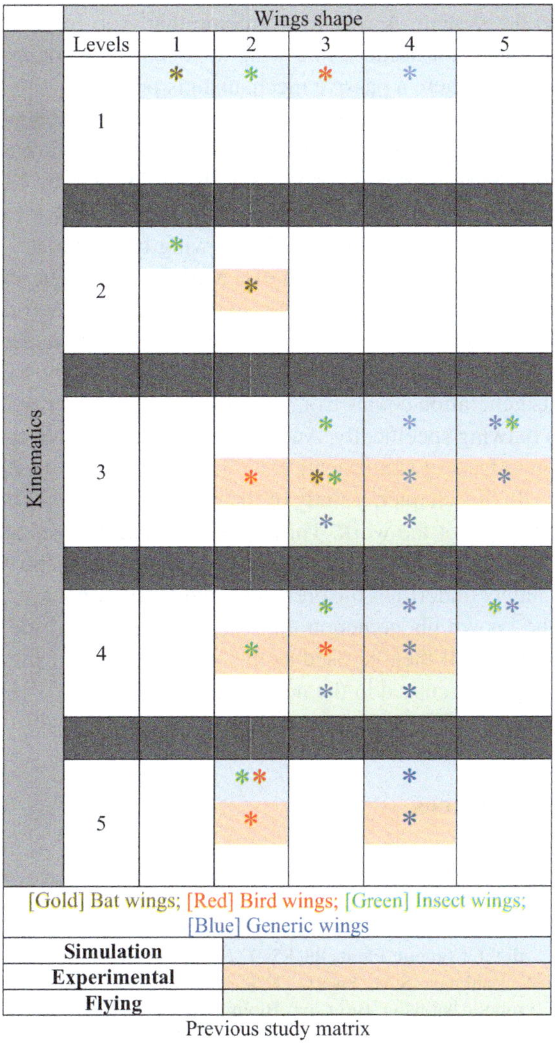

Kinematics \ Levels	Wings shape 1	2	3	4	5
1		* (gold)	* (gold)	* (green)	* (blue)
2		* (green)	* (red)			
3			* (red)	* (green) ** (green) * (green)	* (blue) * (red) * (green)	** (green) * (blue)
4			* (red)	* (green) * (red) * (green)	* (blue) * (red) * (green)	** (green)
5			** (green) * (red)		* (blue) * (red)	

[Gold] Bat wings; [Red] Bird wings; [Green] Insect wings; [Blue] Generic wings

Simulation	(blue)
Experimental	(orange)
Flying	(green)

Previous study matrix

be simplified. One of the gaps found in the review, it is found that there is a jump in simplification between a bat margin shape and the generic wing shape. What is lost between the simplification process is yet to be discovered. In terms of wing kinematics, the wing kinematics that is closest to an in-vivo batwing but still able to produce a flying device is the combination of flapping and twisting combination and wing kinematics at this level is promising. However, observations were done in other studies suggest that wing area changing plays a bigger role since it is the one that morphs the wing the most and plays the biggest role in overall lift generation. A key to the simplification process lies in minimising the number of active actuators that are involved in the mechanism since it adds to the weight and power requirements to the system. A way to overcome that is to introduce a passive mechanism. This is important because in a wing area changing kinematics wing bending is the only motion where a passive mechanism is possible.

In terms of the aerodynamic performance and phenomenon, the review shows that one of the main factors that affect the wing performance is the ability for the wing to produce vortexes particularly the WTV and the LEV. The stronger and longer a wing can generate a vortex and not letting the vortex to shed away from the wing surface, the more overall lift the wing can generate. The key to understanding the aerodynamic phenomenon of a MAV wing design will lie in understanding how to control the vortex generation so that favourable aerodynamic performance could be achieved. While several works have been done in understanding the vortex generation of a batwing, much of the knowledge effects of the LEV and the wing tip vortex and its generation comes from works done using other types of the wing and not done on a batwing specifically. Another knowledge gap that can be found is in understanding the effects of simplification on the aerodynamic phenomenon and performance.

In this chapter, a study of the previous works has been done to establish the basic concepts of the work. This chapter also underlines all the natural observations that have been done in bats which include the wing shape and wing kinematics. Also, this chapter underlines the previous methods of mimicking the natural batwing, underline the known phenomenon that was discovered and the knowledge gap that exist. All of this will then be used as the starting point for the approach of this book, which will be discussed in the next chapter.

References

D.L. Altshuler, R. Dudley, C.P. Ellington, Aerodynamic forces of revolving hummingbird wings and wing models. J. Zool. **264**(4), 327–332 (2004). https://doi.org/10.1017/S0952836904005813

H. Aono, H. Liu, Flapping wing aerodynamics of a numerical biological flyer model in hovering flight. Comput. Fluids **85**, 85–92 (2013). https://doi.org/10.1016/j.compfluid.2012.10.019

J.W. Bahlman, S.M. Swartz, K.S. Breuer, Design and characterization of a multi-articulated robotic bat wing. Bioinspir. Biomim. **8**(1), 16009 (2013). https://doi.org/10.1088/1748-3182/8/1/016009

J.W. Bahlman, S.M. Swartz, K.S. Breuer, How wing kinematics affect power requirements and aerodynamic force production in a robotic bat wing. Bioinspir. Biomim. **9**(2), 025008 (2014). https://doi.org/10.1088/1748-3182/9/2/025008

D. Bie, D. Li, J. Xiang, H. Li, Z. Kan, Y. Sun, Design, aerodynamic analysis and test flight of a bat-inspired tailless flapping wing unmanned aerial vehicle. Aerosp. Sci. Technol. **112**, 106557 (2021). https://doi.org/10.1016/j.ast.2021.106557

R. Bleischwitz, R. de Kat, B. Ganapathisubramani, Aeromechanics of membrane and rigid wings in and out of ground-effect at moderate Reynolds numbers. J. Fluids Struct. **62**, 318–331 (2016). https://doi.org/10.1016/j.jfluidstructs.2016.02.005

R. Bleischwitz, R. de Kat, B. Ganapathisubramani, On the fluid-structure interaction of flexible membrane wings for MAVs in and out of ground-effect. J. Fluids Struct. **70**, 214–234 (2017). https://doi.org/10.1016/j.jfluidstructs.2016.12.001

F. Bouard, T. Jardin, L. David, Aerodynamics of flapping wings with passive and active deformation. J. Fluids Struct. **128**, 104139 (2024). https://doi.org/10.1016/j.jfluidstructs.2024.104139

References

Y. Chen, J. Zhan, J. Wu, J. Wu, A fully-activated flapping foil in wind gust: energy harvesting performance investigation. Ocean Eng. **138**, 112–122 (2017)

Z. Chen, J. Xu, B. Liu, Y. Zhang, J. Wu, Structural integrity analysis of transmission structure in flapping-wing micro aerial vehicle via 3D printing. Eng. Fail. Anal. **96**, 18–30 (2019). https://doi.org/10.1016/J.ENGFAILANAL.2018.09.017

J.A. Cheney, N. Konow, K.M. Middleton, K.S. Breuer, T.J. Roberts, E.L. Giblin, S.M. Swartz, Membrane muscle function in the compliant wings of bats. Bioinspir. Biomim. **9**(2), 025007 (2014). https://doi.org/10.1088/1748-3182/9/2/025007

X. Cheng, S. Lan, Effects of chordwise flexibility on the aerodynamic performance of a 3D flapping wing. J. Bionic Eng. **12**, 432–442 (2015). https://doi.org/10.1016/S1672-6529(14)60134-7

D.J. Cleaver, D.E. Calderon, Z. Wang, I. Gursul, Lift enhancement through flexibility of plunging wings at low Reynolds numbers. J. Fluids Struct. (2016). https://doi.org/10.1016/j.jfluidstructs.2016.04.004

J. Colorado, A. Barrientos, C. Rossi, K.S. Breuer, Biomechanics of smart wings in a bat robot: morphing wings using SMA actuators. Bioinspir. Biomim. **036006**(7), 16 (2012). https://doi.org/10.1088/1748-3182/8/1/019501

L. Cong, B. Teng, L. Chen, W. Bai, R. Jin, B. Chen, Aerodynamic performance of low aspect-ratio flapping wing with active wing-chord adjustment. J. Fluids Struct. **112**, 103964 (2023). https://doi.org/10.1016/j.jfluidstructs.2023.103964

A. De Rosis, On the dynamics of a tandem of asynchronous flapping wings: lattice Boltzmann-immersed boundary simulations. Phys. A **410**, 276–286 (2014). https://doi.org/10.1016/j.physa.2014.05.041

A. De Rosis, G. Falcucci, S. Ubertini, F. Ubertini, Aeroelastic study of flexible flapping wings by a coupled lattice Boltzmann-finite element approach with immersed boundary method. J. Fluids Struct. **49**, 516–533 (2014). https://doi.org/10.1016/j.jfluidstructs.2014.05.010

S. Deng, M. Percin, B. van Oudheusden, B. Remes, H. Bijl, Experimental investigation on the aerodynamics of a bio-inspired flexible flapping wing micro air vehicle. Int. J. Micro Air Veh. **6**(2), 045002 (2014). https://doi.org/10.1260/1756-8293.6.2.105

X. Dong, D. Li, J. Xiang, Z. Wang, Design and experimental study of a new flapping wing rotor micro aerial vehicle. Chin J Aeronaut (2020). https://doi.org/10.1016/j.cja.2020.04.024

L. Du, X. Sun, Effect of flapping frequency on aerodynamics of wing in freely hovering flight. Comput. Fluids **117**, 79–87 (2015). https://doi.org/10.1016/j.compfluid.2015.05.004

B. Duan, C. Gup, T. Mao, H. Liu, Wing geometry and kinematic parameters optimization of bat-like robot fixed-altitude flight for minimum energy. Aerosp. Sci. Technol. **140**, 108482 (2023). https://doi.org/10.1016/j.ast.2023.108482

I. Fenercioglu, O. Cetiner, Effect of unequal flapping frequencies on flow structures. Aerosp. Sci. Technol. **35**(1), 39–53 (2014). https://doi.org/10.1016/j.ast.2014.02.007

W. Geissler, B.G. van der Wall, Dynamic stall control on flapping wing airfoils. Aerosp. Sci. Technol. (2017). https://doi.org/10.1016/j.ast.2016.12.008

C. Gong, J. Han, Z. Yuan, Z. Fang, G. Chen, Numerical investigation of the effects of different parameters on the thrust performance of three dimensional flapping wings. Aerosp. Sci. Technol. **84**, 431–445 (2019). https://doi.org/10.1016/J.AST.2018.10.021

S. Guo, H. Li, C. Zhou, Y.L. Zhang, Y. He, J.H. Wu, Analysis and experiment of a bio-inspired flyable micro flapping wing rotor. Aerosp. Sci. Technol. **79**, 506–517 (2018). https://doi.org/10.1016/J.AST.2018.06.009

W. He, X. Mu, Y. Chen, X. He, Y. Yu, Modeling and vibration control of the flapping-wing robotic aircraft with output constraint. J. Sound Vib. **423**, 472–483 (2018). https://doi.org/10.1016/J.JSV.2017.12.027

S. Heathcote, Z. Wang, I. Gursul, Effect of spanwise flexibility on flapping wing propulsion. J. Fluids Struct. **24**(2), 183–199 (2008). https://doi.org/10.1016/j.jfluidstructs.2007.08.003

C.M. Hoke, J. Young, J.C.S. Lai, Effects of time-varying camber deformation on flapping foil propulsion and power extraction. J. Fluids Struct. **56**, 152–176 (2015). https://doi.org/10.1016/j.jfluidstructs.2015.05.001

T.Y. Hubel, N.I. Hristov, S.M. Swartz, K.S. Breuer, Changes in kinematics and aerodynamics over a range of speeds in Tadarida brasiliensis, the Brazilian free-tailed bat. J. R. Soc. Interface. **9**(71), 1120–1130 (2012). https://doi.org/10.1098/rsif.2011.0838

T.Y. Hubel, N.I. Hristov, S.M. Swartz, K.S. Breuer, Time-resolved wake structure and kinematics of bat flight. Anim Locomot pp. 371–381 (2010). https://doi.org/10.1007/978-3-642-11633-9_29

F.J. Huera-Huarte, M. Gharib, On the effects of tip deflection in flapping propulsion. J. Fluids Struct. **71**, 217–233 (2017). https://doi.org/10.1016/j.jfluidstructs.2017.04.003

H. Isakhani, S. Yue, C. Xiong, W. Chen, X. Sun, T. Liu, Fabrication and mechanical analysis of bioinspired gliding-optimized wing prototypes for micro aerial vehicles, in *2020 5th International Conference on Advanced Robotics and Mechatronics (ICARM)* (2020), pp. 602–608. https://doi.org/10.1109/ICARM49381.2020.9195392

M. Jankauski, Z. Guo, I.Y. Shen, (2018). The effect of structural deformation on flapping wing energetics. J. Sound. Vib. 429, 176–192 (2018). https://doi.org/10.1016/J.JSV.2018.05.005

G. Jeanmonod, M. Olivier, Effects of chordwise flexibility on 2D flapping foils used as an energy extraction device (2017). https://doi.org/10.1016/j.jfluidstructs.2017.01.009

V. Joshi, R. Jaiman, G. Li, Full-scale aeroelastic simulations of hovering bat flight, in *AiAA Scitech 202 Forum*. Orlando, Florida (2020), p. 0335. https://doi.org/10.2514/6.2020-0335

E.K.V. Kalko, Insect pursuit, prey capture and echolocation in pipistrelle bats (Microchirptera) 861–880 (1995)

Z. Kan, Z. Yao, D. Li, D. Bie, Z. Wang, H. Li, J. Xiang, Design and flight test of the fixed-flapping hybrid morphing wing aerial vehicle. Aerosp. Sci. Technol. **143**, 108705 (2023). https://doi.org/10.1016/j.ast.2023.108705

M. La Mantia, P. Dabnichki, Effect of the wing shape on the thrust of flapping wing. Appl. Math. Model. **35**(10), 4979–4990 (2011). https://doi.org/10.1016/j.apm.2011.04.003

R. Lahoti, A. Gogulapati, P. Ghandhi, Design and development of a folding mechanism for bat-like bioinspired wing. IFAC PapersOnLine **55**(22), 400–405 (2022). ISSN 2405-8963, https://doi.org/10.1016/j.ifacol.2023.03.067.

T.Q. Le, J.H. Ko, D. Byun, S.H. Park, H.C. Park, Effect of chord flexure on aerodynamic performance of a flapping wing. J. Bionic Eng. **7**(1), 87–94 (2010). https://doi.org/10.1016/S1672-6529(09)60196-7

Y.J. Lee, K.B. Lua, T.T. Lim, K.S. Yeo, A quasi-steady aerodynamic model for flapping flight with improved adaptability. Bioinspiration Biomim. **11**(3), 036005 (2016). https://doi.org/10.1088/1748-3190/11/3/036005

H. Li, S. Guo, Y.L. Zhang, C. Zhou, J.H. Wu, Unsteady aerodynamic and optimal kinematic analysis of a micro flapping wing rotor. Aerosp. Sci. Technol. (2016). https://doi.org/10.1016/j.ast.2016.12.025

G. Li, Y.Z. Law, R.K. Jaiman, A novel 3D variational aeroelastic framework for flexible multibody dynamics: application to bat-like flapping dynamics. Comput. Fluids **180**, 96–116 (2018). https://doi.org/10.1016/J.COMPFLUID.2018.11.013

K. Liu, S. Bifeng, C. Ang, W. Zhihe, X. Dong, Y. Wenqing, Effects of dynamical spanwise retraction and stretch on flapping-wing forward flights. Chin J Aeronaut **37**(4), 181–202 (2024). ISSN 1000-9361; https://doi.org/10.1016/j.cja.2024.01.006

K.B. Lua, S.M. Dash, T.T. Lim, K.S. Yeo, On the thrust performance of a flapping two-dimensional elliptic airfoil in a forward flight. J. Fluids Struct. (2016). https://doi.org/10.1016/j.jfluidstructs.2016.07.012

X. Meng, M. Sun, Wing kinematics, aerodynamic forces and vortex-wake structures in fruit-flies in forward flight. J. Bionic Eng. **13**, 478–490 (2016). https://doi.org/10.1016/S1672-6529(16)60321-9

M. Moriche, O. Flores, M. García-Villalba, Three-dimensional instabilities in the wake of a flapping wing at low Reynolds number. Int. J. Heat Fluid Flow **62**, 44–55 (2016). https://doi.org/10.1016/j.ijheatfluidflow.2016.06.015

References

F.T. Muijres, P. Henningsson, M. Stuiver, A. Hedenström, Aerodynamic flight performance in flap-gliding birds and bats. J. Theor. Biol. **306**, 120–128 (2012). https://doi.org/10.1016/j.jtbi.2012.04.014

M. Olivier, G. Dumas, A parametric investigation of the propulsion of 2D chordwise-flexible flapping wings at low Reynolds number using numerical simulations. J. Fluids Struct. (2016). https://doi.org/10.1016/j.jfluidstructs.2016.03.010

M. Olivier, G. Dumas, Effects of mass and chordwise flexibility on 2D self-propelled flapping wings. J. Fluids Struct. (2016). https://doi.org/10.1016/j.jfluidstructs.2016.04.002

J. Pfliger, C. Breitsamter, Gust response of an elasto-flexible morphing wing using fluid-structure interaction simulations. Chin. J. Aeronaut. **37**(2), 45–57 (2024). https://doi.org/10.1016/j.cja.2023.12.017

T.N. Pornsin-Sirirak, S.W. Lee, H. Nassef, J. Grasmeyer, Y.C. Tai, C.M. Ho, M. Keennon, MEMS wing technology for a battery-powered ornithopter, in *Proceedings IEEE Thirteenth Annual International Conference on Micro Electro Mechanical Systems (Cat. No.00CH36308)*, vol. 043(2) (2000), pp. 799–804. https://doi.org/10.1109/MEMSYS.2000.838620

T.N. Pornsin-Sirirak, Y.-C. Tai, C.-M. Ho, M. Keennon, Microbat: a palm-sized electrically powered ornithopter, in *Proceedings of NASA/JPL Workshop on Biomorphic Robotics* (2001), pp. 14–17, http://ho.seas.ucla.edu/wp-content/uploads/2011/04/jpl10_2001.pdf

A. Rahman, D. Tafti, The role of vortex–vortex interactions in thrust production for a plunging flat plate. J. Fluids Struct. **96**, 103011 (2020). https://doi.org/10.1016/j.jfluidstructs.2020.103011

D.K. Riskin, D.J. Willis, J. Iriarte-Díaz, T.L. Hedrick, M. Kostandov, J. Chen, D.H. Laidlaw, K.S. Breuer, S.M. Swartz, Quantifying the complexity of bat wing kinematics. J. Theor. Biol. **254**(3), 604–615 (2008). https://doi.org/10.1016/j.jtbi.2008.06.011

D.K. Riskin, J.W. Bahlman, T.Y. Hubel, J.M. Ratcliffe, T.H. Kunz, S.M. Swartz, Bats go head-under-heels: the biomechanics of landing on a ceiling. J. Exp. Biol. **212**(Pt 7), 945–953 (2009). https://doi.org/10.1242/jeb.026161

W. Shyy, Y. Lian, J. Tang, D. Vieru, H. Liu, *Aerodynamics of Low Reynolds Number Flyers* (Cambridge University Press, Cambridge, 2008). https://doi.org/10.1017/CBO9781107415324.004

A.J. Skulborstad, Y. Wang, J.D. Davidson, S.M. Swartz, N.C. Goulbourne, Polarized image correlation for large deformation fiber kinematics. Exp. Mech. **53**(8), 1405–1413 (2013). https://doi.org/10.1007/s11340-013-9751-4

A. Song, X. Tian, E. Israeli, R. Galvao, K. Bishop, S. Swartz, K. Breuer, Aeromechanics of membrane wings with implications for animal flight. AIAA J. **46**(8), 2096–2106 (2008). https://doi.org/10.2514/1.36694

S. Sterbing-D'Angelo, M. Chadha, C. Chiu, B. Falk, W. Xian, J. Barcelo, J.M. Zook, C.F. Moss, Bat wing sensors support flight control. Proc. Natl. Acad. Sci. U.S.A. **108**(27), 11291–11296 (2011). https://doi.org/10.1073/pnas.1018740108

A.K. Stowers, D. Lentink, Folding in and out: passive morphing in flapping wings. Bioinspir. Biomim. **10**, 025001 (2015). https://doi.org/10.1088/1748-3190/10/2/025001

X. Sun, J. Zhang, Effect of the reinforced leading or trailling edge on the aerodynamic performance of a perimeter-reinforced membrane wing. J. Fluids Struct. **68**, 90–112 (2017)

W.B. Tay, Symmetrical and non-symmetrical 3D wing deformation of flapping micro aerial vehicles. Aerosp. Sci. Technol. (2016). https://doi.org/10.1016/j.ast.2016.05.026

W.B. Tay, J.H.S. de Baar, M. Percin, S. Deng, B.W. van Oudheusden, Numerical simulation of a flapping micro aerial vehicle through wing deformation capture. AIAA J. **56**, 3257–3270 (2018). https://doi.org/10.2514/1.J056482

X. Tian, J. Iriarte-Diaz, K. Middleton, R. Galvao, E. Israeli, A. Roemer, A. Sullivan, A. Song, S. Swartz, K. Breuer, Direct measurements of the kinematics and dynamics of bat flight. Bioinspiration & Biomim. **1**(4), S10–S18 (2006). https://doi.org/10.1088/1748-3182/1/4/S02

F.B. Tian, H. Luo, J. Song, X.Y. Lu, Force production and asymmetric deformation of a flexible flapping wing in forward flight. J. Fluids Struct. **36**, 149–161 (2013). https://doi.org/10.1016/j.jfluidstructs.2012.07.006

A.J. Torregrsosa, A. Gil, P. Quintero, A. Cremades, Multi-fidelity approach to the numerical aeroelastic simulation of flexible membrane wings. Aerosp. Sci. Technol. **155**, 109673 (2024). https://doi.org/10.1016/j.ast.2024.109673

R.M. Waldman, A.J. Song, D.K. Riskin, S.M. Swartz, K.S. Breuer, P. Researcher, Aerodynamic behavior of compliant membranes as related to bat flight, June 2008 (2008), pp. 1–13

S. Wang, X. Zhang, G. He, T. Liu, Lift enhancement by dynamically changing wingspan in forward flapping flight. Phys. Fluids **26**, 061903 (2014). https://doi.org/10.1063/1.4884130

J. Wang, C. Zhang, Z. Wu, J. Wharton, L. Ren, Numerical study on reduction of aerodynamic noise around an airfoil with biomimetic structures. J. Sound Vib. **394**, 46–58 (2017). https://doi.org/10.1016/j.jsv.2016.11.021

Q. Wen, S. Guo, H. Li, W. Dong, Nonlinear dynamics of a flapping rotary wing: modeling and optimal wing kinematic analysis. Chin. J. Aeronaut. **31**(5), 1041–1052 (2018). https://doi.org/10.1016/J.CJA.2018.03.004

Y. Wenqing, W. Liguang, X. Dong, S. Bifeng, Aerodynamic performance of micro flexible flapping wing by numerical simulation. Procedia Eng. **99**, 1506–1513 (2015). https://doi.org/10.1016/j.proeng.2014.12.702

G.M. Whitesides, Bioinspiration: something for everyone (2015). https://doi.org/10.1098/rsfs.2015.0031

J. Wu, S.C. Yang, C. Shu, N. Zhao, W.W. Yan, Ground effect on the power extraction performance of a flapping wing biomimetic energy generator. J. Fluids Struct. **54**, 247–262 (2015a). https://doi.org/10.1016/j.jfluidstructs.2014.10.018

J. Wu, J. Wu, F.B. Tian, N. Zhao, Y.D. Li, How a flexible tail improves the power extraction efficiency of a semi-activated flapping foil system: a numerical study. J. Fluids Struct. **54**, 886–899 (2015b). https://doi.org/10.1016/j.jfluidstructs.2015.02.006

J. Wu, J. Qiu, Y. Zhang, Automated kinematics measurement and aerodynamics of a bioinspired flapping rotary wing. J. Bionic Eng. **14**(4), 726–737 (2017). https://doi.org/10.1016/S1672-6529(16)60438-9

J. Xiang, J. Du, D. Li, K. Liu, Aerodynamic performance of the locust wing in gliding mode at low Reynolds number. J. Bionic Eng. (2016). https://doi.org/10.1016/S1672-6529(16)60298-6

W. Yang, L. Wang, B. Song, Dove: a biomimetic flapping-wing micro air vehicle. Int. J. Micro Air Veh. **10**(1), 70–84 (2018). https://doi.org/10.1177/1756829317734837

D. Yin, Z. Zhang, The inertial power and inertial force of robotic and natural bat wing. Comptes Rendus - Mecanique **344**, 195–207 (2016). https://doi.org/10.1016/j.crme.2015.11.002

Y. Yu, Z. Guan, Learning from bat: aerodynamics of actively morphing wing. Theor. Appl. Mech. Lett. **5**, 13–15 (2015). https://doi.org/10.1016/j.taml.2015.01.009

H. Yusoff, M.Z. Abdullah, K.A. Ahmad, M.K. Abdullah, S. Suhaimi, Experimental study on the effect of skin flexibility on aerodynamic performance of flapping wings for micro air vehicles. Appl. Mech. Mater. **629**, 18–23 (2014). https://doi.org/10.4028/www.scientific.net/AMM.629.18

H. Yusoff, M.Z. Abdullah, M. Abdul Mujeebu, K.A. Ahmad, Experimental study on the effect of skin flexibility on aerodynamic performance of flexible skin flapping wings for micro air vehicles. Exp. Tech. **39**(1), 11–20 (2015). https://doi.org/10.1111/ext.12004

Z. Zhang, X. Sun, P. Du, J. Sun, Y. Wu, Design of a hydraulically-driven bionic folding wing. J. Mech. Behav. Biomed. Mater. (2018). https://doi.org/10.1016/j.jmbbm.2018.03.024

M. Zhang, B. Song, X. Yang, X. Lang, J. Xuan, L. Wang, Analysis of the integrated pattern of hoverable flapping wing micro-air vehicle. J. Eng. Res. (2024). https://doi.org/10.1016/j.jer.2024.06.007

Chapter 3
Bat-Inspired Wing Design

Abstract In this chapter, the approach of the book is outlined where it begins with the method used for wing generation and then followed by the flight conditions used. It is then followed by the simulation approach used and the test procedures that were used. The study will end with the final semi-active mechanism study that was done.

As mentioned before, the focus of this book can be divided into two. The first part focuses on the bio-inspiration problem of translating observation found in nature into solutions for a mechanical problem. The second part focuses on solving the aerodynamic problem of designing a flapping wing that can be used for a flapping MAV. Therefore, the approach that was used for this study revolved around fulfilling these two parts. To solve the design problem, a MAV wing geometry and wing kinematics were designed that are based on a generalised and averaged observation of a bat. For the wing to be possible for MAV use, the wing geometry and kinematics were simplified. Also, to base the design of the wing on an in vivo observation, the intended operating condition of the design was based on the known operating condition of a bat. To solve the aerodynamic part of the study, the simplified wing design was tested for its aerodynamic performance (lift and drag) and aerodynamic efficiency (lift over drag). A flow field testing was also conducted to observe the key vortices that were involved to produce a favourable wing design. At the end of the study, a wing design that can be used for a flyable MAV design.

There are three objectives outlined in this book. The first objective is to produce a wing design based on the kinematics and geometry of a batwing. The second is to study the aerodynamic performance and efficiency of a bio-inspired batwing. The third is to propose a design that can be used for a flying MAV wing. The flow of the study in this book must therefore reflect the objectives of the book.

The first objective was achieved by the first couple of major steps done in this study which is the background research and the wing geometry design. The result of the background research has been outlined in the previous chapter, but it is mentioned again here because a part of bio-inspired design is the design based on natural observations. Therefore, the background research was done to find an averaged observation of a natural batwing, to find the observation of the wing kinematics, and to find the observed flying conditions of a batwing. Once the information has been gathered, the next step is to design a wing geometry that is based on batwing and simplified designs of that wing geometry. The information from the background research will also be used to determine the test condition for this study.

The second objective was achieved by testing the aerodynamic effects of the designed wing designs. This was done by doing two types of studies simultaneously: a test study and a simulation study. The test work was done by first designing a flapper mechanism that can produce the required flapping motion. The wing was then tested in an open-circuit wind tunnel to test the aerodynamic performance and efficiency by measuring the lift and drag and by calculating the lift over the drag of the wing. The simulation study was done to observe the airflow pattern generated by the wing and to understand the vortices that play a role in producing the observed aerodynamic performance. This is done by simulating the flapping wings to match the same conditions that have been observed in the test study. If the calculated aerodynamic performance and efficiency of the wing were observed, then it can be inferred that the calculated flow pattern is valid. If the calculation does not fit the observed test results, then the work must be repeated so that validating matches can be achieved.

The third objective is achieved by comparing the results gathered during the previous steps of the study. The goal of this book is to find the best wing design that can be used for a flying MAV device. Therefore, a right combination of the best wing geometry and the best bending flexibility that produces a wing that has the best aerodynamic performance and efficiency was chosen as the proposed wing for a flying MAV. Figure 3.1 shows a process flow chart of the wing geometry generation process.

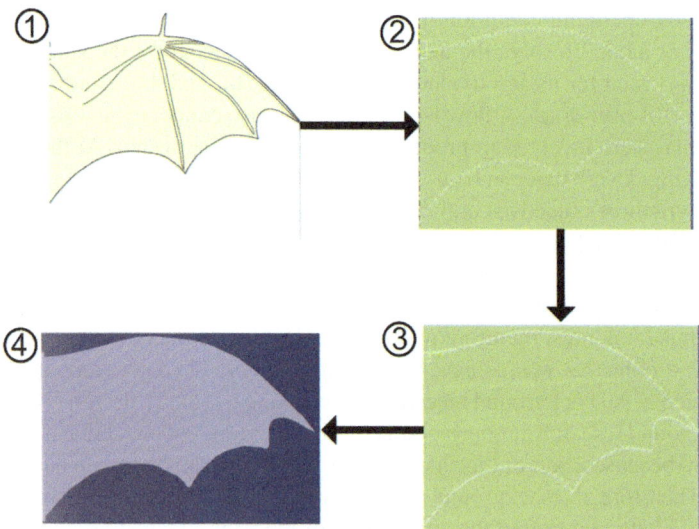

Fig. 3.1 Flow chart of the wing geometry generation process

3 Bat-Inspired Wing Design

Several challenges need to be solved when it comes to bio-inspiration design. As mentioned before, these challenges come from translating the natural observations into the mechanical applications. Nature is overly complex with no apparent standards or convention and nature is highly diverse. The diversity of nature is not only coming from a large number of known bat species but also within the same species, there are no two individual bats have the same wing dimensions or the same wing mass.

This book overcame these problems by applying several measures. The problem of diversity was overcome by using a generalised and averaged observation of a batwing as its source of wing shape inspiration. The wing shape used by this book was an averaged observation of a bat species *Cynopterus brachyotis* with its wings extended. However, only the wing shape and the wing proportions were kept but the exact dimensions of the wing itself were changed. This is because the wing size needed to fit into the design constraints of a MAV design which requires the wingspan to be 150 mm or less.

Once the wing shape observation was chosen the next step that needed to be completed was translating the observed wing shape to produce a wing geometry to be used for a MAV design. The problem of complexity can be solved by simplifying the geometry of the wing itself. As mentioned before, the highest level of simplification where an example of a flying device can be found is a wing that has the margin or outline shape of a batwing but the wing itself is a uniform flat wing. This wing model is similar to a work done by Ishimoto and Sugimura (2017), where the margin wing shape of insects was studied.

This book achieves this by taking the chosen wing observation mentioned above and transferring it into a Computer-Aided Design (CAD) software, CATIA V.6 and used the image as a reference for generating the wing geometry. To this study, the wings were done separately and without the body. This was done because this book ignores the presence of the bat body for both its test and simulation tests since the body shape is unlikely to be used in a real MAV design. The wings are modelled separately because they needed to be fabricated separately for the test study and the simulation test will be done using a half body model, both subjects will be covered later in the book. Because the wings needed to be modelled separately, the wing stretch was 60 mm long and the rest of the wing dimensions is scaled to fit this dimensional restriction. The reason the wing stretch was 60 mm long is that this allows for the test wing model to fit into the 150 mm MAV design constraint. The wing reference image was then sectioned into 50-line sections all at equal distance from the root of the wing to the wingtip. The lines were used to generate vertices that will later be used as a building block of the wing model by generating a vertex where the line section meets the outline of the wing. This was done for all line sections and both at the leading edge and the trailing edge of the wing. At the end of the process, 99 vertices were generated. The next step was using the generated vertices to produce an outline of the batwing by connecting the generated vertices using a spinal line. The final step of the wing geometry generation process was simply padding out the wing outline to produce a flat wing with a batwing margin that has a thickness of 1 mm.

Fig. 3.2 Final wing model

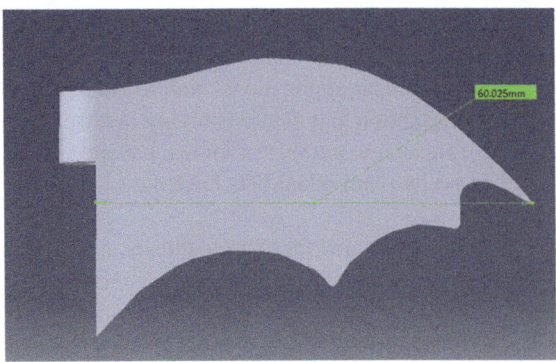

Table 3.1 Final wing model dimensions

Dimensions	Values
Wing surface area	13 cm^2
Chord length	34 mm
Wing thickness	1 mm
Root to tip length	60 mm

It is important to note that the method used to generate the wing geometry is not a unique process. The method was widely used as a reverse engineering process that allows for 2D drawings or image of a design to be transferred into a 3D CAD drawing. The uniqueness of the method described in this book is its usage in a Bio-Inspiration design context. The result of the process is a wing geometry that will be used for later tests. The wing geometry and its dimensions can be seen in the Fig. 3.2 and Table 3.1.

One of the advantages of the method used in this book is that it allows for the wing margin shape itself to be further simplified. As mentioned before in the background research done for this book, one of the gaps that can be found in wing simplification methods is the huge gaps that exist between a wing margin and a generic wing where the wing margin is a wing shape is based on or somewhat based on an in vivo batwing, while the generic wing is not based on any natural wing at all. Currently, there are no methods for abstracting a wing shape that is based on a natural wing. However, there are methods of geometry simplifications that have been done in other fields particularly in the field of geometry laser scanning. The basic concept of geometry simplification is done by skipping over the building block vertices that make up a certain geometry and the result is a simplified geometry. An example of this process can be found in a work done by Xiong et al. (2016).

In this book, this method was done by removing every other the vertices that were used as the building block to generate the wing. The reason 99 vertices were used to generate the original wing because it was enough the capture the complete features of the natural wing. The wing geometry was simplified by removing every other vertex and using the remaining vertices to produce a new wing geometry. The wing can

3 Bat-Inspired Wing Design

then be further simplified by removing more and more vertex until, when there are five vertices available to draw a wing margin, the final product is a generic elliptical wing. Five wings were used for this book, 99 vertices, 49 vertices, 29 vertices, 21 vertices, and 5 vertices wing. Ninety-nine and 49 vertices wings are chosen because the shape of the wings closely mimics the geometry of a natural wing. five vertices wing was chosen because the final product was a generic elliptical wing. The 29 and the 21 vertices wings were chosen because the produced shapes fall somewhere in between natural wings and a generic elliptical wing. The different generated wings can be seen in Table 3.2.

Another advantage of this method is that the different wing geometries have the same wing area which is 13 cm^2. This is crucial because once the wing surface area plays a role in lift generation. Having the wing area have the same surface area means that any difference in aerodynamic performance is only due to the wing geometry and not due to different wing geometry.

Just like the generation of the wing geometry, the condition for the wing design must also follow a natural observation. To this book, the flight condition for the wing refers to the flight condition that the wing will be tested under. For this book, the flight conditions parameters that taken into consideration were the flight speed, the flapping frequency, the flapping angle, and the test angles of attack, AoA.

Therefore, this book chose 8.5 Hz as the test flapping frequency. With the flight speed of the wing was chosen at 4 m/s and the flapping frequency was chosen to be 8.5 Hz. The relation between the flapping frequency can be expressed by the reduced frequency, and it can be expressed by

$$k = \frac{\omega c}{2U_{\text{ref}}}$$

where ω is the angular velocity of a flapping wing, c is the wing's reference chord, and U_{ref} is the forward flight velocity. The test flapping frequency that was chosen means that the angular velocity, ω, was 53.4 rad/s. As mentioned before, the reference chord, c, was 34 mm, and the forward flight speed, U_{ref}, was 4 m/s. This means that the test reduced frequency was 0.2 which makes the test flight condition to fall into the unsteady flight regime.

Two types of angles are concerned when it comes to a flapping wing. The first is the angle of attack, AoA, where the test AoA for this book is from 0° to 45° to capture crucial features such as the stall angle. The negative angle of attack was not tested for this book because flapping wings do not have a zero-lift angle of attack because of the constant thrust generated by the wing motion, the wing does not produce a negative lift. The AoA is also chosen to be higher because of one of the advantages of a flapping wing especially one that is based on a batwing that has a delayed stall angle. The other type of angle that is concerned this book is the flapping angle which can be defined as the total travel angle during the flapping motion. This book opted to choose the flapping angle at 55°

These flight conditions were chosen to reflect the nature of bio-inspiration design since most of the parameters were based on in vivo observations of a batwing. This

Table 3.2 List of all generated wings

Wing name	Geometry
99 vertices	
49 vertices	
29 vertices	
21 vertices	
5 vertices	

3 Bat-Inspired Wing Design

Table 3.3 Flight conditions value

Item	Values
Flight speed	4 m/s
Air density	1.225 kg/m^3
Flapping angle	55°
Flapping frequency	8.5 Hz
Angle of attack	0°–45°

also will give a picture of the possible flight capabilities of the future MAV design. Therefore, the chosen flight condition parameter values were used for both test and simulation tests. Full values of the flight condition parameter tests can be seen in the Table 3.3.

As mentioned before, the first adjustable factor that was tested was the effects of geometry wing. This was done using both test and simulation methods. The simulation study was done to produce the airflow visualisation results for the study while the test study was done to verify the global value results of the simulation study. While the test and the simulation study were two different processes and each has its steps, it is interesting to note that each of these studies was done at almost concomitantly. This is done so that one study does not significantly inform the other and the agreement between the results arrives independently.

The test process begins with the design and the fabrication of the flapper and wing system. The design was done using the Computer-Aided Design (CAD) software, CATIA to design the system. The flapper system used a brushless DC motor with a series of gears and trusses to produce the required movement. The wings and flapper parts were fabricated using 3D print with Poly-lactic Acid (PLA) as a material.

Once the fabrication was done, the next step was testing the aerodynamic performance of the fabricated wing design. The aerodynamic test was done using an open circuit subsonic wind tunnel. The aerodynamic performance was tested by measuring the lift and the drag generated by the wing. These generated forces were measured using a series of a strain gauge attached on a series of parallelograms that are attached to a Data Acquisition System (DAQ) that is attached to a personal computer. Each of the wing designs was tested at angles of attack from 0° to 45° with 5° increments. This means that the tested angles of attack that were tested for each of the wing designs are 0°, 5°, 10°, 15°, 20°, 25°, 30°, 35°, 40°, 45°.

After the lift and drag data was collected, the recorded data was then processed using a data processing software, Kyowa PCD-30. The data processing was achieved by filtering the mechanical and electronic noise using Butterworth filtration equation. The filtered data were then used to calculate the aerodynamic performance by calculating the lift coefficient, C_L, and drag coefficient, C_D. The aerodynamic efficiency was then calculated by calculating the C_L/C_D.

The first step of the test study is the design of the flapper system and the fabrication of the fabrication and the test wing geometry. The purpose of the flapper system for this book is to fulfil the stated flight kinematics parameters, especially when it comes to flapping angle and flapping frequency. To achieve the required wing motion, the

Fig. 3.3 Flapper used in the study

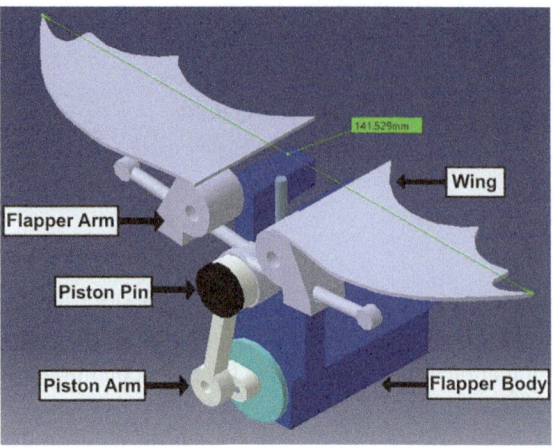

flapper is designed several gears (motor gear and follower gear) and trusses (the body, the slider, the follower, and the arm). The main purpose of the body is to house the main mechanism, to provide a platform to attach the mechanism to the gauge system, and to provide a pivot point for the wings to be able to execute the flapping motion. The purpose of the follower and the slider was to transfer the circular motion of the main gear system into a cyclical vertical system which is transferred into a flapping motion for the wing by the arms of the flapper system. The design of the wing geometry was adjusted by adding a truss follower that allows it to be attached to the body of the flapper system and attached to the arm of the flapper system that drives the wing and producing the wing-flapping motion. The dimensions of the flapper system were designed to allow for the system to be able to produce a flapping motion of the wing that has a symmetrical 55° flapping angle (Fig. 3.3).

The work required for the flapper system was done using a 12 V brushless DC motor that has a motor gear attached and used to drive the follower gear which was attached to the follower truss. The motor was powered by a transformer power supply that allows for adjustment of the supply voltage which is important for reaching the required flapping frequency. The flapping frequency was determined by the speed of the DC motor which was determined by the supply voltage. The flapping frequency was measured using two methods, the first method is by using the force gauge and counting the number of cycles recorded over time. The second method is by using a high-speed camera and visually counting the number of cycles produced over time. The two methods are used to validate each other and measuring the flapping frequency are important since the speed control of the system is an open-loop system.

For the flapper system, the motor, gear motor, follower gear, the screws, and the pins were all standard part. The rest of the fabrication system was fabricated by using the 3D printing method called Fluid Deposition Modelling (FDM) or sometimes referred to as Fluid Filament Fabrication (FFF) which is an additive manufacturing process that builds an object by selectively depositing thermoplastic polymers that usually comes in a filament form in a predetermined path layer-by-layer. The

3 Bat-Inspired Wing Design

Table 3.4 Mechanical properties of PLA

Item	Values
Tensile strength	37 MPa
Elongation	6%
Flexural modulus	4 GPa
Density	1.3 g/cm^3
Melting point	60 °C

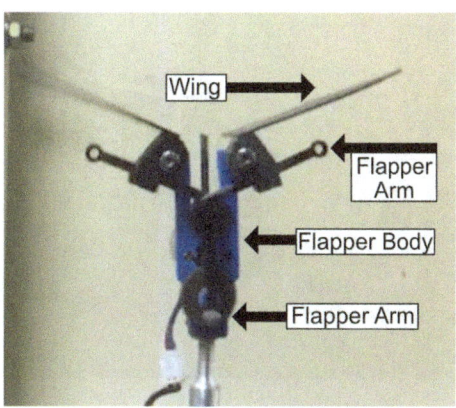

Fig. 3.4 Fabricated flapper using 3D-printing

thermoplastic polymer material that was used in this book was PLA. The FDM 3D printing method was used for this study because the method was relatively quick and economical. However, the final product does have a drawback of having a rough surface layer. This was not a hindrance for this book because the main source of drag for a flapping wing is not really friction drag but instead is induced drag and this means that the rough surface is less of a factor. Plus, since all the wings had the relatively same surface roughness, the surface roughness did not have an overall effect on the final results. The material PLA was chosen because of it is very economical, the material was rigid enough that it minimises flexure during testing, and the material was light enough that it did not produce too much mechanical noise that cannot be filtered out. The specifications of the fabricated material can be seen in the Table 3.4 (Fig. 3.4).

The test largely was done in an open-loop sub-sonic wind tunnel specifically the wind tunnel that was available in the Fluid Mechanics Laboratory in the School of Aerospace Engineering, University Sains Malaysia, Engineering Campus. The purpose of the wind tunnel was to provide the required wind conditions that match the set flight condition for the designed wing. The wind tunnel that was used as an open-loop or open return wind tunnel which means that the air that was blown into the test section of the wind tunnel was gathered from the surrounding area instead of from the internal circuit of the wind tunnel itself. The required wing speed was achieved by the blower section of the wing tunnel and then assumed to be laminar

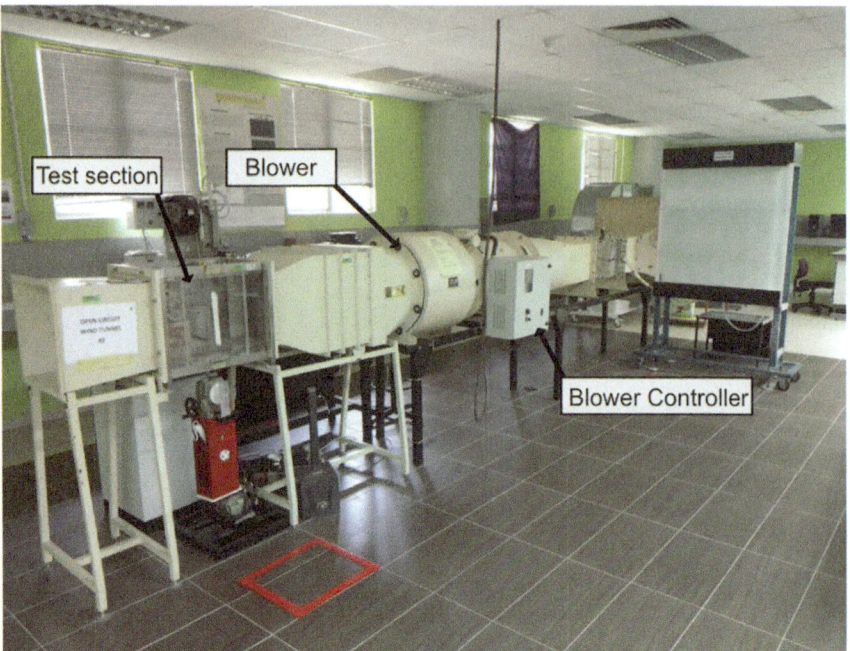

Fig. 3.5 The sub-sonic open channel wind tunnel

and uniform by the straightener or the honeycomb section of the wind tunnel. The speed of the blower, which produces the speed of the wind, was controlled by an open-loop feedback controller that changes the frequency of the blower itself rather than the resulting wind directly. The wind speed was therefore controlled by setting the blower speed that was prescribed by the wind tunnel's manual and by validating with a measurement done using an anemometer. Figure 3.5 shows the wind tunnel that was used in this study and the sections that were involved.

This study uses a time-averaged lift and drag instead of the raw instantaneous lift and drag because the purpose of the book is to study the global aerodynamic forces instead of the transitional lift and drag. Once the time-averaged lift and drag were calculated, the lift coefficient, $C_{L\ Avg}$, and the drag coefficient, $C_{D\ Avg}$, were calculated by using the formula:

$$C_{LAvg} = \frac{L_{avg}}{\rho V^2 A / 2}$$

$$C_{DAvg} = \frac{D_{avg}}{\rho V^2 A / 2}$$

This study uses a sampling rate of 1,000 Hz and the measurements were taken for 10 s. This means that the test system samples 10,000 readings for each measurement.

An average of the data set was then taken to produce the L_{avg} and D_{avg}. Lift coefficient and drag coefficient were used instead of time-averaged lift and drag because of the objective of the book was to propose a wing design of a bio-inspired wing for a MAV and a non-dimensional measurement of the lift and drag are more useful for design purposes.

Once the aerodynamic performance of the wings was obtained, it is now possible to calculate the aerodynamic efficiency of the wing this was done by finding the ratio of the generated lift to the generated drag. The measurement, processing, and calculation of the aerodynamic performance (lift and drag) and the aerodynamic efficiency (lift over drag) was done for all wing designs at angles of attack from 0° to 45° at an increment of 5°.

As mentioned before, the test study was done at almost the same time as the simulation study. This was done so that one study was not used to largely inform the other. While this is true, the test study and the simulation study did not occur 100% simultaneously. This is because of the difference between procedures and time frames between the two studies were different. For example, a test study of a single wing design takes approximately 1 day to complete, while the same wing design requires several days for a simulation study.

In this book, the main purpose of a simulation study was to study the flow pattern that was affected by the different wing designs. The reason simulation approach was used for this book was because test flow visualisation approach was unavailable for this study due to lack of resources. For the simulated flow pattern to be valid, the simulation results must agree with the test result within a small margin of error. For this study, the measured parameter that was used to compare the test study and the simulation study was the generated lift and drag (or to be specific, the time-averaged lift, $C_{L\ Avg}$, and the time-averaged drag, $C_{D\ Avg}$). If the simulation results were validated by the test results, then it can be inferred that the simulated flow patterns are the same as the airflow pattern in the test study.

This study used the software ANSYS V16, which was a multi-tool simulation software that can be used for mechanical, electronic, vibration, and fluid flow applications. The software was chosen because of its ability to integrate several simulation modules in one place. The main method of simulation that was used in the simulation study was a method known as a one-way Fluid–Structure Interaction (FSI). FSI is a combination of simulation modules that was used to solve a problem that involves the interaction between the fluid and a given structure. In the software, a native module called 'Systems Coupling' was used to achieve the FSI simulation process by connecting different simulation modules for the simulation process. In this study, the modules used were the 'Transient Structural' and the 'Fluent' structure. The transient structural module was used to define the motion of the wing over time. The fluent module, on the other hand, was used to solve the fluid dynamic aspect of the simulation. The system coupling module transferred the wing motion data in the transient structural module to the fluent module that afterwards calculates the fluid motion and the resulting forces.

Like the test study, the simulation study also measures the wings from 0° to 45° angles of attack. However, for the simulation study, the increments of the angles of attack were at 15° instead of 5°. This means that the angles of attack studied for the simulation were 0°, 15°, 30°, and 45°. This is because the time to solve the simulation was far longer than the time to finish the test study and using a larger increment of the angle of attack for the simulation study was more economical.

The transient structure module that was used is a Finite Element Analysis (FEA) solver. However, in this study, the transient structure module was used to define and model the motion of the wing that occurs over time and no actual FEA simulation has been done. This is because the wing was considered as a single stiff body that does not flex due to the aerodynamic forces and no strain or stress calculation was done on the wing itself. The wing kinematics that was focused on during this phase was the pure flapping kinematics and the main parameter that was being studied was the effect of different wing geometry.

The main governing equation that was used in the fluent module was a modified pressure-based Navier–Stokes equation with transient time function. Pressure-based means that the equation considered that the density of the fluid as unchanging. This assumption was made because of the flight speed of the test was sub-sonic.

The enhanced treatment was used for the turbulence model because of the coarse mesh that was used to reduce the probability of the negative volume mesh error to occur. The enhanced wall treatment worked by blending the wall function by the use of a transition damping function. This was done so that the transition of the different wall functions was smoother. Transient time function means that the calculation was done for fluid that moved over time. The governing equation can be expressed as

$$\rho \left(\frac{\partial u}{\partial t} + u\frac{\partial u}{\partial x} + v\frac{\partial u}{\partial y} + w\frac{\partial u}{\partial z} \right) = -\frac{\partial p}{\partial x} + \mu \left(\frac{\partial^2 u}{\partial x^2} + \frac{\partial^2 u}{\partial y^2} + \frac{\partial^2 u}{\partial z^2} \right) + \rho g_x$$

$$\rho \left(\frac{\partial v}{\partial t} + u\frac{\partial v}{\partial x} + v\frac{\partial v}{\partial y} + w\frac{\partial v}{\partial z} \right) = -\frac{\partial p}{\partial y} + \mu \left(\frac{\partial^2 v}{\partial x^2} + \frac{\partial^2 v}{\partial y^2} + \frac{\partial^2 v}{\partial z^2} \right) + \rho g_y$$

$$\rho \left(\frac{\partial w}{\partial t} + u\frac{\partial w}{\partial x} + v\frac{\partial w}{\partial y} + w\frac{\partial w}{\partial z} \right) = -\frac{\partial p}{\partial z} + \mu \left(\frac{\partial^2 w}{\partial x^2} + \frac{\partial^2 w}{\partial y^2} + \frac{\partial^2 w}{\partial z^2} \right) + \rho g_z$$

$$\frac{\partial u}{\partial x} + \frac{\partial v}{\partial y} + \frac{\partial w}{\partial z} = 0$$

The other equation that was involved in the fluent module calculation was involved with the turbulence model. The turbulence model that was used in this study was the k-epsilon turbulence model with enhanced wall treatment. K-epsilon model was because of the equation was tailored for recirculating flows which were important to capture the important LEV and WTV. The k-epsilon contained two parts; the first part is the equation for the turbulent kinetic energy (k) and the second part is the equation for the dissipation (ε).

$$\frac{\partial}{\partial t}(\rho k) + \frac{\partial}{\partial x_i}(\rho k u_i) = \frac{\partial}{\partial x_j}\left[\left(\mu + \frac{\mu_t}{\sigma_k} \right) \frac{\partial k}{\partial x_j} \right] + P_k + P_b - \rho\epsilon - Y_M + S_k$$

$$\frac{\partial}{\partial t}(\rho \epsilon) + \frac{\partial}{\partial x_i}(\rho \epsilon u_i) = \frac{\partial}{\partial x_j}\left[\left(\mu + \frac{\mu_t}{\sigma_\epsilon}\right)\frac{\partial \epsilon}{\partial x_j}\right] + C_{1\epsilon}\frac{\epsilon}{k}(P_k + C_{3\epsilon}P_b) - C_{2\epsilon}\rho\frac{\epsilon^2}{k} + S_\epsilon$$

The blending function of the enhance wall treatment used can be expressed as

$$u^+ = e^\Gamma u^+_{\text{lam}} + e^{\frac{1}{\Gamma}} u^+_{\text{turb}}$$

$$\Gamma = -\frac{0.01(y^+)^4}{1 + 5y^+}$$

For the results to confirm its accuracy, a validation process is needed. However, because of the relative newness of the bio-inspired FW-MAV field, there are no standards that can be used as a reference for validation. Also problematic is the topic of the book itself is too specific that there are no previous works that can be used as a reference to compare the values of the results. Therefore, a different validation method is needed. In this book, the results of the book were done in two ways.

The first method was an internal validation method where two approaches were done and the results of the two methods were then compared. In this book, this was done for the stiff wing study or the study for the effect of wing geometry where the wings were tested using both test and simulation method. If the results for both of the tests agree with one another within a margin of 10%, then the final results will be considered validated.

The second method of validation is by comparing the final results with previous work. While the specific value of the results cannot be validated by the previous work, the generally observed pattern of the results, however, can be validated. If the observed phenomena found in the results of the book is confirmed by a previous study, then the results will then be considered validated. In this study, there are two places where this validation method was done, the first is after the study of the effect of wing geometry and the second is after the study of the effect of flexible mechanisms.

The final objective of this book is to propose a feasible design for a flying FW-MAV. However, the scope of the study does not include the detailed design and fabrication of an actual flying FW-MAV, this book included an analysis of the final wing configuration and how the tested wing itself can be applied in a future FW-MAV. The analysis uses a preliminary design approach for a UAV outlined by Ong et al. (2019).

In this book, the design process that was done was limited to only the initial concept and design step. The analysis that was done started with the purpose of the FW-MAV and the design requirements of the FW-MAV design was outlined by outlining the possible flying profile and flying environment that the FW-MAV could operate in. Then, an initial concept and design of the FW-MAV were done by outlining the features and restrictions that could be possible of the FW-MAV design.

The approach of the study has three main phases that correlate to the three main objectives. The first phase was the design and fabrication of the wing themselves where observation of a natural wing was used to determine and derive the geometry and the flapping flight condition of the tested wing. The geometry of the tested wing was derived from the wing shape observed of a *Cynopterus brachyotis* bat and from its shape, five wing shape was derived. In the second phase, the lift over the drag of the wing was tested using a simulation study and validated by the test results. The final phase was the evaluation of the design. The outcome of the study will be a wing design strategy for an FW-MAV based on the lessons learned from the measured global values like lift, drag, and lift over drag ratio, and the local wind flow visualisation that explains these values, which will be discussed in the next chapter.

References

Y. Ishimoto, K. Sugimura, A mechanical model for diversified insect wing margin shapes. J. Theor. Biol. Theor. Biol. **427**, 17–27 (2017). https://doi.org/10.1016/J.JTBI.2017.05.026

W. Ong, S. Srigrarom, H. Hesse, Design methodology for heavy-lift unmanned aerial vehicles with coaxial design methodology for heavy-lift unmanned aerial vehicles with coaxial rotors, January 2019 (2019). https://doi.org/10.2514/6.2019-2095

B. Xiong, S.O. Elberink, G. Vosselman, Footprint map partitioning using airborne laser scanning data, in *ISPRS Annals of the Photogrammetry, Remote Sensing and Spatial Information Sciences. Prague, Czech Republic: XXIII ISPRS Congress* (2016), pp. 241–247. https://doi.org/10.5194/isprs-annals-III-3-241-2016

Chapter 4
Bat-Inspired Wing Performance

Abstract In this chapter, the result of the study will be outlined that begins with the aerodynamic result found in all the different wing geometries. This is then followed by the overall comparisons between all the wing geometries. This chapter will end with the results for the wings that include the semi-active mechanism and the difference that the machine has made.

To keep the results of the study that was done for this book coherent, it is important to note the main objective of the study which was, to develop an approach that allows for the translation of natural observation to mechanical applications, to understand the aerodynamic phenomenon of the of a flapping-wing design derived from a batwing and provide a probable design suggestion for a flying Micro Air Vehicle. The first objective was partially by the development of wing derivation method explained in the chapter before. To validate the usefulness of the method, the aerodynamic performance and aerodynamic efficiency were measured and compared. The second objective of the study was achieved by studying the airflow pattern generated by the wing, especially by studying the generated LEV and the generated Wingtip Vortex (WTV). The final objective of the book was achieved by comparing the overall performance of the different wing designs and determining which could be feasible for a flying MAV wing design.

In this chapter, the results of this book study will be discussed that begins with studying the aerodynamic performance and efficiency of each wing and comparing the simulation results and the test results to see the validation of the two methods. After that, the aerodynamic performance and efficiency of wings will be compared and determine which wing geometry has the overall aerodynamic advantage. Once done, the results of the visualisation will be presented and will be used to tally the airflow pattern with the aerodynamic performance. The next step was presenting the results for the wing semi-passive folding mechanism which begins with the selection of the best wing geometry and presenting the aerodynamic performance and efficiency of the wing with the addition of different flexibility. The effects of wing flexibility will be presented by comparing their aerodynamic performance and

© The Author(s), under exclusive license to Springer Nature Singapore Pte Ltd. 2025
S. B. Suhaimi et al., *Flapping Wing Micro Air Vehicles*,
SpringerBriefs in Applied Sciences and Technology,
https://doi.org/10.1007/978-981-96-2908-4_4

efficiency. The airflow pattern will then be presented to show the relationship between the airflow pattern and the aerodynamic performance.

Global Values Comparison Between Wing Geometry

To identify the best wing geometry that has the best aerodynamic performance and efficiency, it is useful to compare each wing side by side. The values that will be used for comparison are the test result values because the test results are more detailed in terms of the global lift, drag, and efficiency values. The aerodynamic performance for the case of the study is defined as the generated time-averaged lift coefficient, $C_{L\,avg}$, and the generated time-averaged drag coefficient, $C_{D\,avg}$. The aerodynamic efficiency on the other hand is defined as the lift to drag ratio, $C_{L\,avg}/C_{D\,avg}$. It is desired for the wing to have the most $C_{L\,avg}$ while generating the least $C_{D\,avg}$. This is because the more lift that was generated measures the better aerodynamic performance while greater drag can contribute to lesser aerodynamic efficiency. As mentioned before, it was shown in previous studies that the main problem in a flapping flight might lie in the generated induced drag that was inherited to the wing type. In this book, the $C_{L\,avg}$, the $C_{D\,avg}$, and the $C_{L\,avg}/C_{D\,avg}$ were measured from angles of attack from $0°$ to $45°$. This was done both in test and in a simulation.

The generated lift is important because it is the desired value when it comes to wing design. It is usually the marker for aerodynamic performance with better lift generated points towards a better aerodynamic performance. The Fig. 4.1 shows the $C_{L\,avg}$ against AoA curve for all the different wings. From the collected data, it can be observed that, at low AoA, the generated lift and generated drag are almost the same for all wing geometry. However, as the AoA increases (at AoA $15°$ and above), the results $C_{L\,avg}$ and the $C_{D\,avg}$ became distinct. In terms of how the wings are different from one another, the results show a difference between the wings that are closer to the natural wing (i.e. the 99 vertices wing, 49 vertices wing, and the 29 vertices wing) and the wings that are closer to the generic wing (i.e. 21 vertices wing and the 5 vertices wing). It can be observed that the generic wings can generate the most lift with 5 vertices wing generating the most $C_{L\,avg}$ followed by the 21 vertices wing, especially at early AoA. The wings that are closer to a natural or in vivo wing are found to have lesser lift generation, especially at lower AoA, with 29 vertices generated the most $C_{L\,avg}$ among them, followed by the 49 vertices wing and the 99 vertices wing.

However, the difference between the two wing types can also be seen in lift slope and stall angles. The wing that is closer to the generic wing is found to have a larger lift slope with the 5 vertices wing having a larger lift slope, followed by the 21 vertices wing. The wings that are closer to natural wings are found to have lower lift slope with 29 vertices wing having the most lift slope among them, followed by the 49 vertices wing and then followed by the 99 vertices wing. In terms of stall angle, however, it can be observed that the wings that are closer to natural wing have larger stall angles than the ones that are close to a generic wing. The 99 vertices wing has the largest stall angle followed by the 49 vertices wing, then followed the 29 vertices wing, the 21 vertices wing, and the 5 vertices wing are found to have the smallest stall angle.

Fig. 4.1 Comparison results for the $C_{L\,avg}$ against AoA

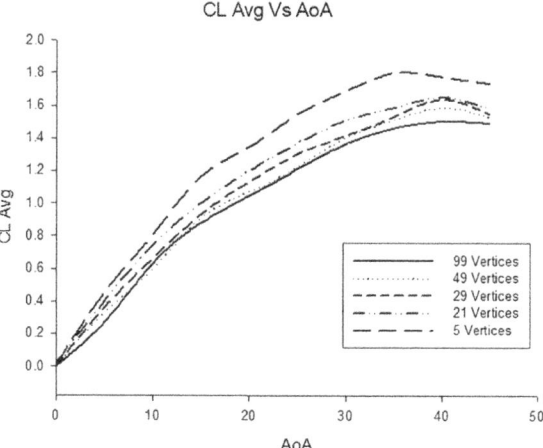

The comparison between results of the lift generated by different wing geometries has shown a clear trade-off as the wing changes from being closely mimicking a natural wing to be a generic elliptical wing. The closer a wing is to be a generic wing, the better the lift generation but this is at the cost of having a lower stall angle. The closer a wing is to mimic a natural wing the lift generation might be lower, but the wing will have an advantage of delayed stall angles. In terms of the comparison between the most conventional wing (the 5 vertices wing) and the closest to the natural wing (99 vertices wing), it can be seen that the lift is lower for the 99 vertices wing by around 25% at earlier AoA and around 15% at later AoA.

In terms of drag, however, the values of drag are usually undesirable because the drag values are detrimental to the aerodynamic efficiency of the wing. According to Sachs (2015), it was found that one of the main concerns for a flapping-wing design is the higher levels of induced drag generated by the flapping wing. The Fig. 4.2 above shows the comparison results for the $C_{D\,avg}$ versus AoA for the different wings. The results show that the wings that are closer to generic wings generated higher levels of drag compared to wings that are closer to natural wings. It can be observed that the 5 vertices wing generates the highest level of drag, followed by the 21 vertices wing, then the 29 vertices wing, the 49 vertices wing, and the 99 vertices wing generated the least drag. This is interesting because, while it is true that the wings that are closer to a natural wing generated less drag, the most modified among the natural wings generated the least level of drag. When comparing the 99 vertices wing with the 5 vertices wing, it can be observed that the 99 vertices wing has a lower drag by 33% at the point with the largest difference (35° AoA).

As mentioned before, the aerodynamic efficiency in this study was measured in terms of $C_{L\,avg}/C_{D\,avg}$ since it calculates the balance between the desired performance values, which is the generated lift, and the cost of the flight, which is the generated drag. In terms of $C_{L\,avg}/C_{D\,avg}$, as can be seen in the Fig. 4.3 it can be observed that the most efficient wings are the ones that are closest to natural wings in most AoA

Fig. 4.2 Comparison results for the $C_{D\,avg}$ against AoA

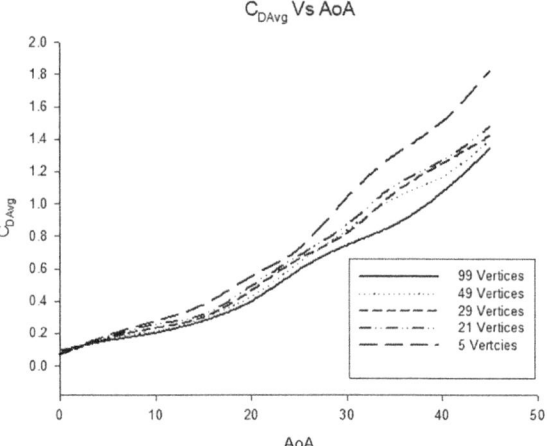

Fig. 4.3 Comparison results for the $C_{L\,avg}/C_{D\,avg}$ against AoA

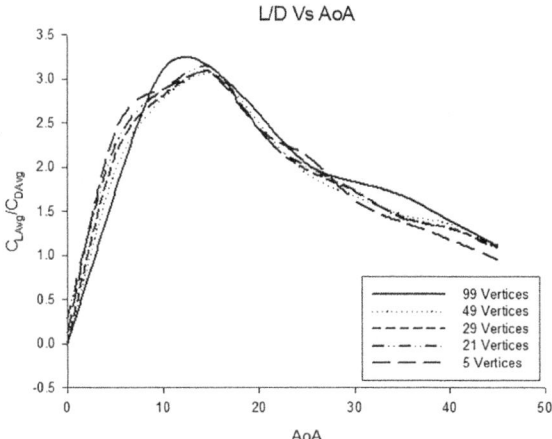

(especially at AoA 10° and above). While the results are not clear for all AoA which wing geometry has the best efficiency, a trend can be observed from the results. The results show that, in most cases, the wing with the lowest efficiency is the 5 vertices wing, followed by the 21 vertices wing, followed by the 29 vertices wing, then the 99 vertices wing, the 49 vertices wing, and the wing with the highest efficiency is the 99 vertices wing beyond 10°.

Overall, the results show a trade-off between wing geometry that mimics the closest to natural wing and the more generic wings. Wings that closely mimic natural wings with the main features of the wing geometry are still recognizable even if the features are changed, that the aerodynamic performance is lesser because of the lesser lift generated by the wings. However, in terms of aerodynamic efficiency, the closer to natural wings are better. The wings also have the distinct advantage of delayed stall angle. The opposite features are also observed with the wings that are more generic and closer to a conventional wing than a natural wing generated larger overall lift with a larger lift slope, but the stall angle happens sooner. More generic wings also sacrifice in terms of aerodynamic efficiency where the $C_{L\ avg}/C_{D\ avg}$ is observed to be lower. Comparison between the 99 vertices wing and the 5 vertices wings has shown that the largest difference shows that 99 vertices show an improved efficiency of 7.3%.

The question that remains is which wing is the best among all the tested wings. The goal of the study is to find a wing that can be used for a flying MAV design that requires the wing to generate enough lift so that the device can fly but have good enough aerodynamic efficiency to help with the limited space for more power supply. After considering all of the design requirements, it can be found that the 99 vertices wing is the best among all other wings since it is the one that the least generated drag among all wing. The wing is also found to have good lift slope and delayed stall angle. The lower lift generation however can be overcoming with the addition of the flexible bending mechanism.

For the visualisation results to be considered as acceptable, an internal validation of the results first needed to be done. This is because the visualisation results were done purely using simulation CFD methods which needed to be based on a real-life test. Therefore, a comparison was done between the global results from the test and the simulation study. Once the comparisons are found to be close, it can be deduced that the local airflow simulation results are reasonably accurate. The Fig. 4.4 shows the $C_{L\ Avg}$, the $C_{D\ Avg}$, and the $C_{L\ Avg} / C_{D\ Avg}$ comparison between the simulation and the test results.

For all of the different wing vertices, the differences between the test and simulation are all below 10%. This means that there is an agreement between the test and simulation results. However, the difference between the test and simulation results is not zero. The difference that exists is due to several reasons. The main reason is due to the assumptions made in the simulation that occurs in during the test procedures. For example, wing flexure does not occur during the simulation process because the wing was considered as a rigid body and it does not flex or bend during the flapping process. However, during the test process, the wing does bend and flex even if it is just at a small magnitude. Another source of the difference between the test and simulation is the errors that are inherently unavoidable such as the overall accuracy of the test apparatus, the outside mechanical and electronic noise that affects the end recorded result, and the mechanical noise from the flapper system itself.

Once the results of the global values for aerodynamic performance and efficiency are obtained, the local values of airflow needed to be examined to explain the observed data found in the global aerodynamic performance and efficiency results. The local examination of the local values is done by examining the visualisation of the airflow using CFD simulation methods. As can be seen above, the difference between the simulation results and the test results for $C_{L\ avg}$, $C_{D\ avg}$, and $C_{L\ avg}/C_{D\ avg}$ is less than 10%. This means that the simulation results are validated, and it can be argued that the visualisation of the airflow done in the CFD calculation is also valid. In this study,

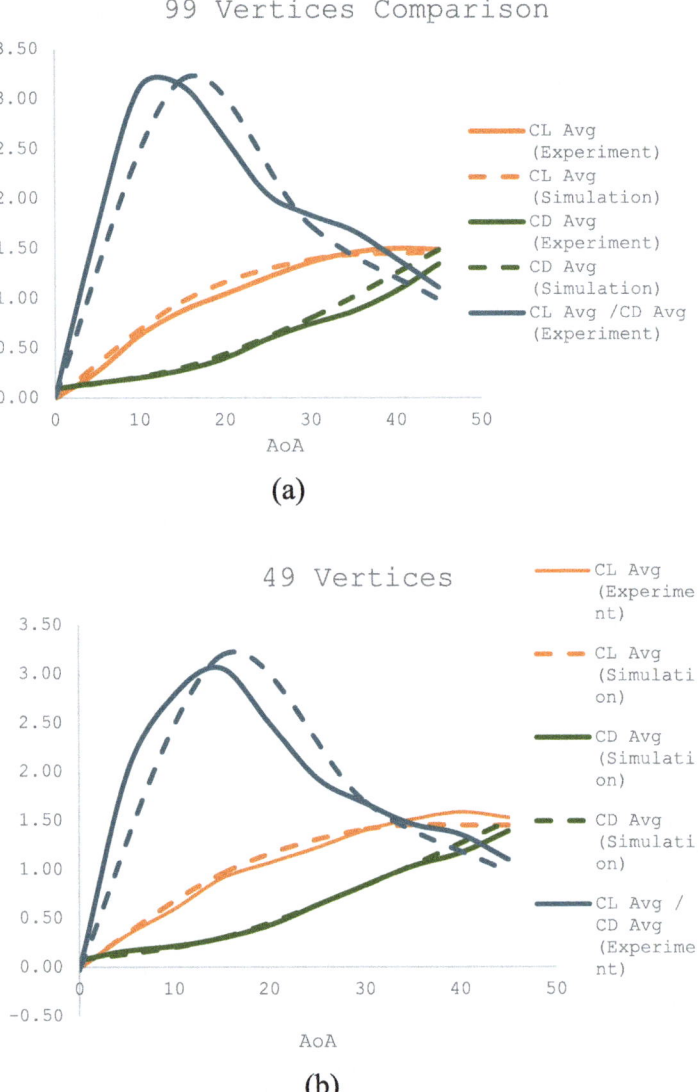

Fig. 4.4 Simulation and test results comparison where; **a** is for the 99 vertices wing, **b** is for the 49 vertices wing, **c** is for the 29 vertices wing, **d** is for the 21 vertices wing, and **e** is for the 5 vertices wing

two visualisation results are examined, the first is the wing motion and the second is the airflow visualisation.

In terms of airflow visualisation, the main aim of the analysis is to study and analyse the presence of the WTV and the LEV. This is because of the role of the two vortices played in the lift generation and could explain the observation that

4 Bat-Inspired Wing Performance

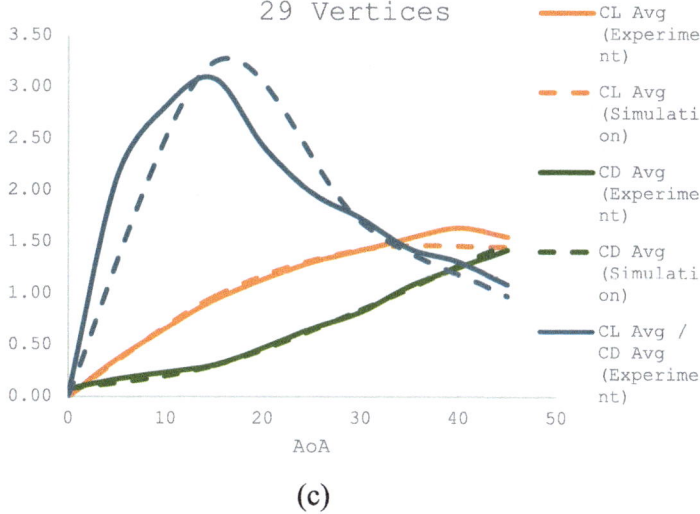

(c)

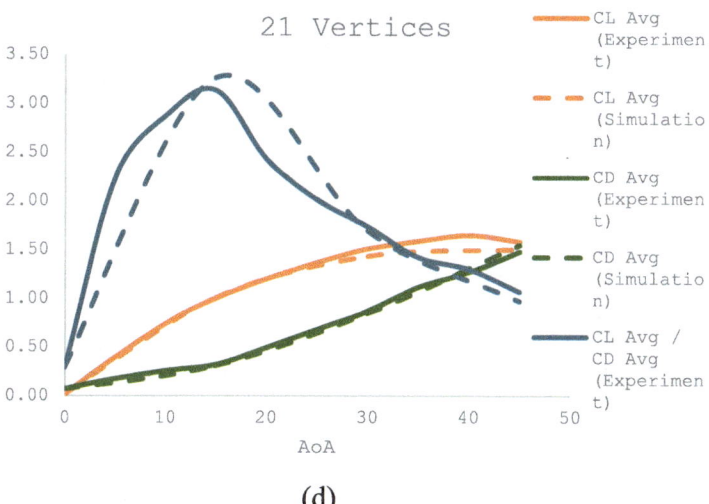

(d)

Fig. 4.4 (continued)

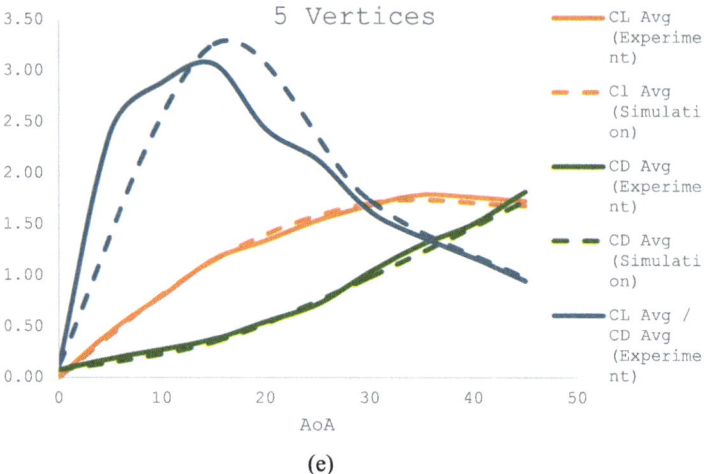

(e)

Fig. 4.4 (continued)

was made in the previous section. The airflow visualisation was represented by the airflow streamline which not only will be useful to look at the size and strength of the generated vortices. The streamline was coloured according to the local air velocity with red being the highest air velocity and blue showing the lowest air velocity. The scale of the local air velocity can be seen near the visualisation result figures. The visualisation result will also be done on all the AoA that was done using the simulation method (i.e. the 0°, 15°, 30°, and 455° AoA). For the sake of clarity, only the crucial visualisation results are shown.

To avoid confusion with a visual representation of the airflow being too busy and analysable, the streamline analysis will be done in a section plane of the wing that highlights the formation of different vortices. For the LEV, the location for the visualisation results will be at 30 mm from the wing root which can be seen in the Fig. 4.5.

LEVs can be defined as the swirling air motion that formed at the leading edge (or the trailing edge of the wing in the case of TEV) due to the motion of the wing and the interaction between the wing's leading edge (and trailing edge) and the incoming free stream air velocity (Fig. 4.6).

LEV forms as the wing moves downward during the downward stroke, the leading edge of the wing pushes the airflow and causes the airflow to spill over from the bottom of the wing to the top of the wing which then will be pushed by the forward-moving freestream and this causes the airflow to curve back downstream. As the wing continues to move down, this will pull that part of the airflow back to the top of the wing surface and this causes the LEV to develop (Fig. 4.7).

The development of LEV over different phases of the wing-flapping motion and different AoA changes on a similar general pattern across all of the wing geometries. The difference between the wing geometries occurs in terms of vortex strength and

4 Bat-Inspired Wing Performance

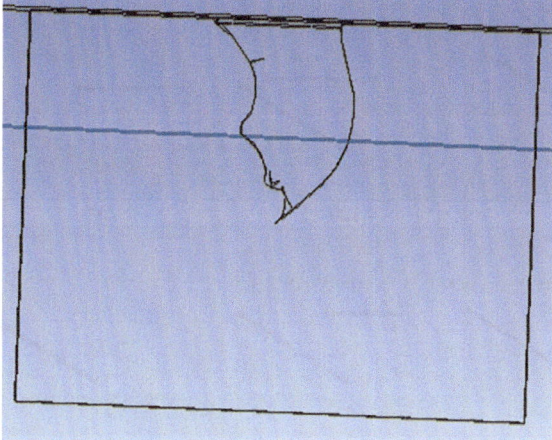

Fig. 4.5 Location of the cross-section plane

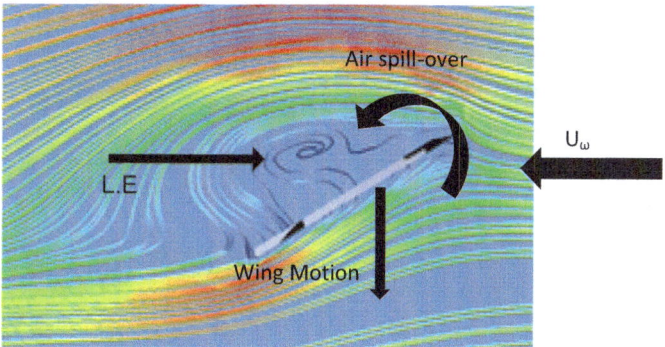

Fig. 4.6 Example of the formation of LEV

at which AoA where certain vortex develops. To better illustrate the development of the LEV, Fig. 4.8 shows shows the velocity vector for the 99 vertices wing at different periods and different AoA. It can be observed that, during the down-stroke motion phase, the LEV will increase in size as the wing continues to move down but as the wing arrives at the low angle or the end of the LEV begins to decrease its size and further on the decrease as the wing motion changes its direction. As the upstroke motion begins, the LEV decays almost immediately until the LEV is undetectable. At the end of the upstroke motion, the LEV has completely gone.

It is also important to note that, as the AoA increases, the size of the LEV also increases and the velocity of the airflow near to the wing also increases with increasing size of the LEV. At larger AoA, however, where the stall angle can be found, it can be observed that there is a second vortex started to appear at the trailing edge of the wing. This vortex is observed to allow for the local airspeed at the trailing edge to

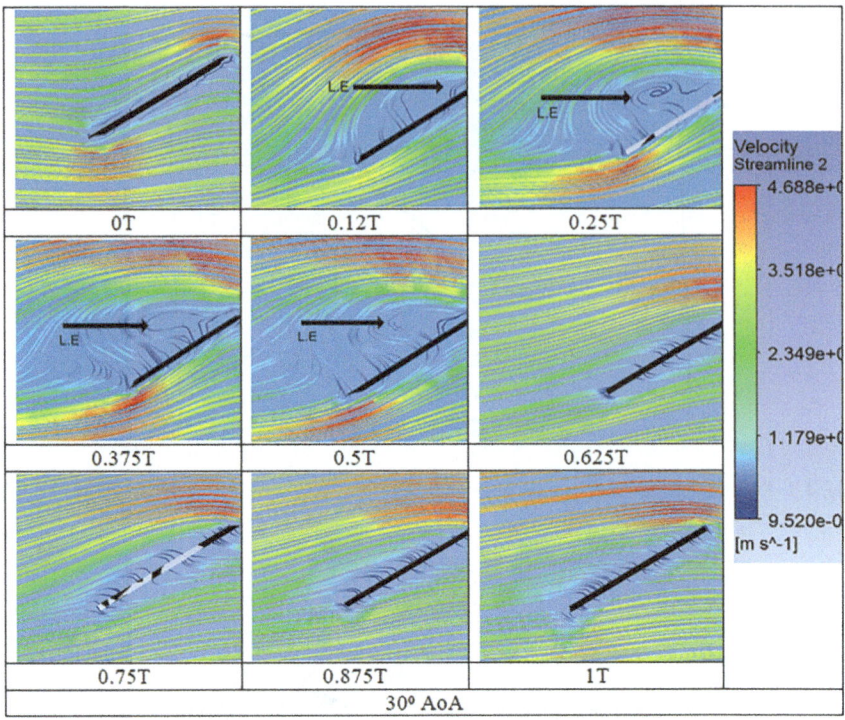

Fig. 4.7 LEV development for 99 vertices wing

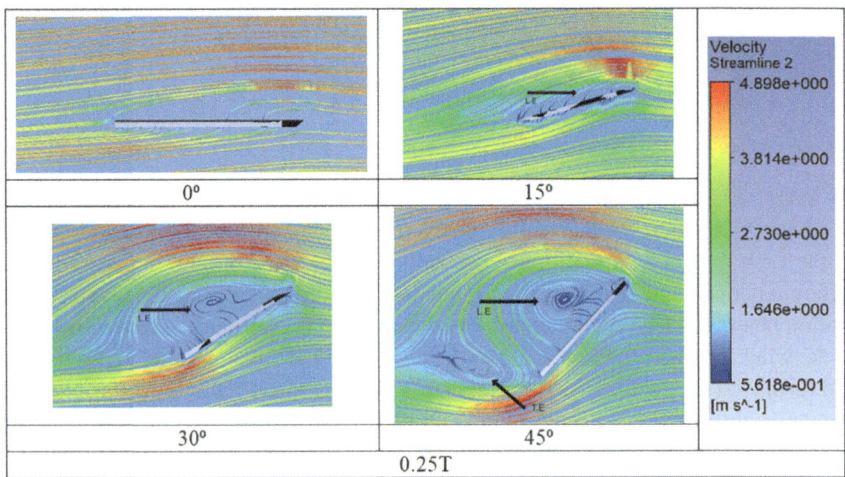

Fig. 4.8 LEV development for different AoA for 99 vertices wing at 0.25 T time phase

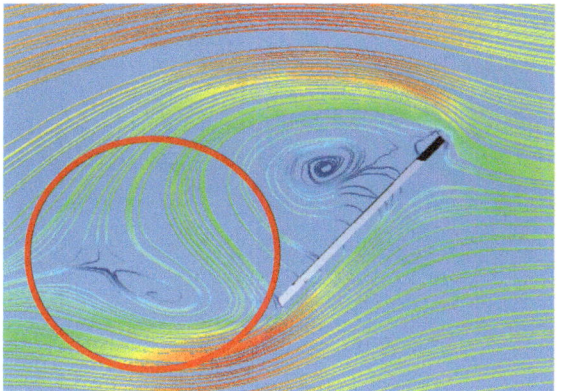

Fig. 4.9 Close up of the TEV

increase which generates a downward force. This then explains the decrease in lift generation at the stall angle (Fig. 4.9).

The relation between the LEV and the generated lift and drag can be explained because LEV pushes the freestream airflow to move away from the surface of the wing and the airflow moves as if the wing itself has a camber. The larger the LEV the further away from the incoming airflow from the wing surface and the larger the velocity of the airflow that moves above the wing which generated larger levels of lift. However, this also causes an increase in the levels of induced drag. This is most obvious with the sharp increase in drag that was recorded at AoA 15° and above which was caused by the large size of the LEV. The size of the leading edge will also be important in explaining the difference between generated lift and drag for different wing geometries.

To compare the airflow pattern of different wing geometries and its effect, the Fig. 4.10 shows the velocity vector for all wing geometries at all phases of the flapping motion for the 30° AoA and at 0.25 T. It is shown only at 30° AoA because at 30° AoA is when the differences in recorded generated lift and drag between the different wing geometries begin to manifest. The reason why the 0.25 T time step was chosen because it is at this time step when the wing is at 0° flapping angle which allows for the vortices to be shown clearly. Plus, a figure that shows all AoA will be too busy and it will be difficult to get a point across but the complete result of all of the other flow visualisations will be shown later in this section.

As mentioned before, the overall pattern of LEV development is similar for all wing geometries. The difference between them is the size of the LEV itself and its consequential effect. From the global value results, it was discovered that the wing geometries that are the closest to natural wings (the 99, 49, and 29 vertices wings) have a lower generated lift but have a lower generated drag and delayed stall. It was also discovered that wing geometries that are the closest to conventional wings (the 21 and 5 vertices wings) generated larger lift but also larger drag and have an earlier AoA.

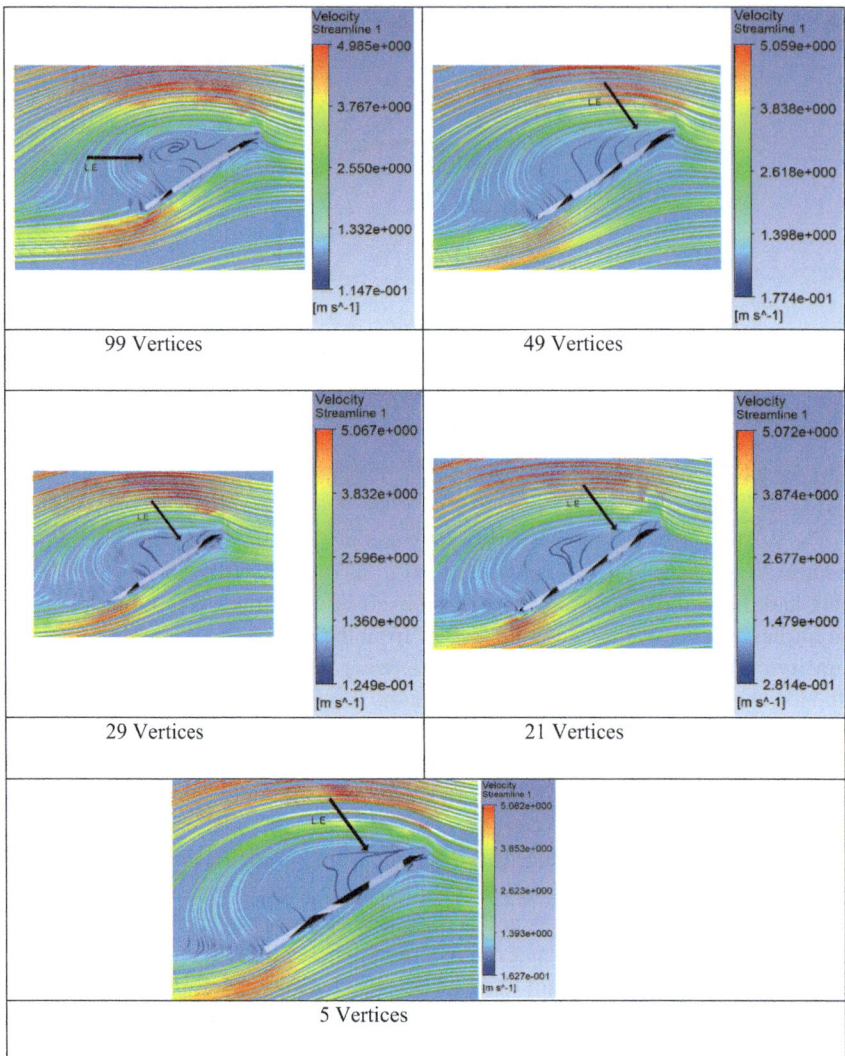

Fig. 4.10 LEV for all wing geometries at 30° AoA at 0.25 T

One clue that points towards the explanation of the difference between the aerodynamic performances of the different wings lies in the way the different wings can be distinguished. Due to the way the wing geometries were generated, the leading edges of the wings are all similar to one another, but the trailing edges of the wings are vastly different. When the wing is closely mimicking a natural wing, the curves of the wings are more pronounced and closely resemble the curves of a natural batwing. However, as the vertices are removed and the wing became closer and closer to a conventional generic wing, the trailing-edge started to straighten out until the elliptical wing-shaped 5 vertices wing, where the trailing edge is completely straight. While the wing geometry generation method allows for the wing surface area and the chord length of the wings were all the same, the distance between the leading edge and the trailing edge of them in different sections of the different wings are all varied.

This is important because the flow visualisation results show that the reason wings that are closer to generic wings have larger lift and drag generation is because of the shape of the trailing edge. If the wing shapes that are at the two extreme were compared (the 99 vertices wing and the 5 vertices wing), it can be observed that the local max velocity for the airflow in the 5 vertices wing is higher than that of the 99 vertices wing. The reason for this observation can be attributed to the fact that the LEV for the 5 vertices wing is stronger than that of the 99 vertices wing. Further investigation made into the flow pattern of these two wings shows that the LEV of the 99 vertices wing spilt over the trailing edge of the 99 vertices wing which caused the LEV to be diminished. This, however, is not seen in the 5 vertices wing where the LEV remains on the surface of the wing (Fig. 4.11).

This difference in trailing edge also can be used to explain why the wings that are closer to generic wings have a quicker stall angle and the wings that are closer to natural wings have a delayed stall angle. This is due to the straightened trailing-edge sections of the wings in the more generic wings that are further away from the leading-edge sections allows for quicker development of the counter Trailing Edge Vortex (TEV) which is the main factor for stall angle.

In terms of drag, the observation found in the results of the global values can be explained by the generation of induced drag, where the more lift generated is also resulted in larger drag. This also explains why wings that larger lift have also had a larger drag. Plus, it can be assumed that other types of drag do not have a larger role because the aerodynamic efficiency of all of the wings is all similar indicating that there are no outside factors that are at play.

One section that is interesting to note is that looking and the velocity vertex for all of the wings at 45° AoA, it can be seen that the TEV for the 5 vertices wing is stronger and more developed than all of the other trailing edge on all of the rest of the wing geometries. This is because, at 45° AoA, the 5 vertices wing is further away from its stall angle compared to the rest of the wings. This validates the observation that the 5 vertices wing geometry has a quicker stall angle compared to all other wings. This also shows that, if the AoA continues, the TEV will continue to grow, causing a downward force and further reducing the generated lift.

In terms of the Wingtip Vortex (WTV), the plane for streamline visualisation is set to be located at 10 mm from the wing's leading edge. This plane allows for the visualisation from the root of the wing to the tip of the wing (Fig. 4.12).

The WTV is defined as the formation of a circulating air motion that formed at the wingtip (or at the wing root for the WRV) due to the wing motion and the result of the interaction between the wing and the airflow. Due to the difference in how the wingtip (and wing root) vortex develops, the visualisation results needed to be laid out differently where the down-stroke motion phase and the upstroke motion

Fig. 4.11 Trailing edge spillover where; **a** is the 99 vertices wing at 30° AoA at 0.25 T and **b** is the 5 vertices wing at 30° AoA at 0.25 T

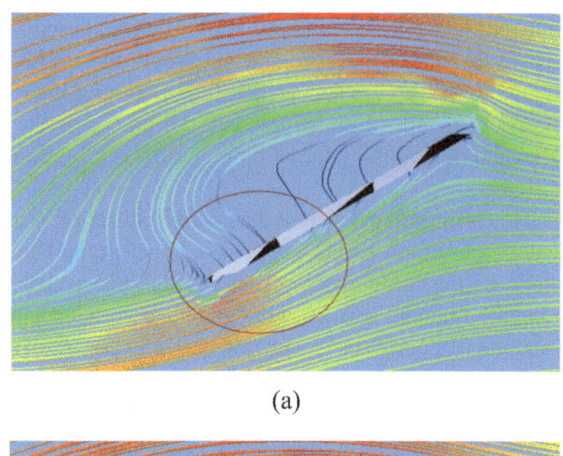

(a)

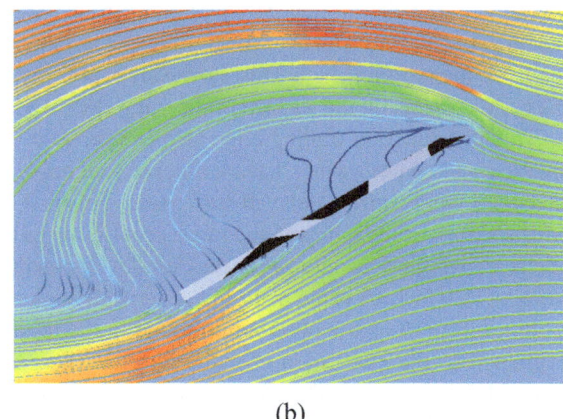

(b)

Fig. 4.12 Wingtip vortex visualisation cross-section plane

4 Bat-Inspired Wing Performance

Fig. 4.13 Example of the formation of WTV

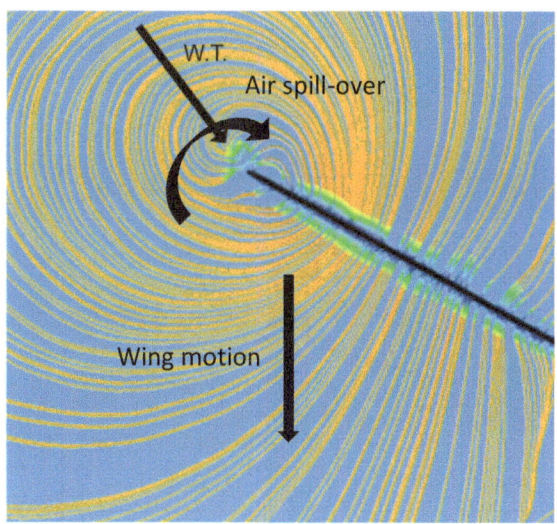

phase needed to be discussed separately. This is because the formation of the WTV is highly dependent on the motion of the wingtip itself. Unlike the LEV, the WTV forms almost entirely since the wingspan of the wing is finite and can never be formed if the wing is assumed as an infinite (or a two-dimensional) wing.

The Fig. 4.13 shows an example of the formation of WTV where, it was observed that as the down-stroke begins, the movement of the wing pushes the air downwards creating a gust of wind velocity that moves in the same direction that the wing is moving. However, as the down-stroke motion continues, the air begins to spill over from the bottom of the wing to the top of the wing. This spillover creates a vortex at the edge of the wing which is the wingtip or the wing root which is present at the top of the wing.

The Fig. 4.14 shows the flow visualisation of the Wingtip Vortex (WTV) and Wing Root Vortex (WRV) development of 99 vertices wing at 0° AoA at all time steps. At 0° AoA, the strongest vortex that can be observed is the WTV. This is because of the angular motion of the wing, there is a larger displacement of the wingtip compared to the wing root that barely moves. This means that more air is pushed at the wingtip, which causes more spillover and causes a stronger vortex. As the wing down-stroke motion continues, the vortex increases in strength but decreases in strength as the wing reaches the end of the wing down-stroke phase due to the angle of the wing. For the upstroke phase, the phase begins as the wing moves up and the WTV that was left during the down-stroke phase begins to die down. As the wing continues to move upwards, the motion will push the air upwards and airflow begins to spill over from the top of the wing surface and into the bottom of the wing surface. This then causes a similar WTV, but the direction of the vortex is the opposite and forms at the bottom of the wing. Also, the WTV grows as the upstroke motion continues

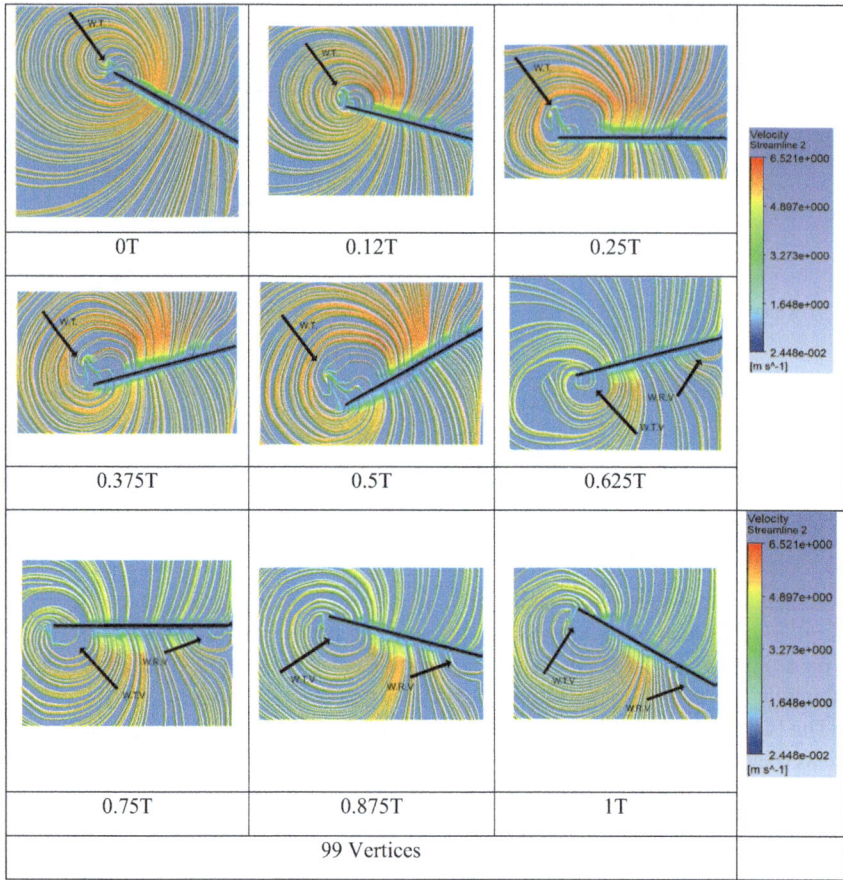

Fig. 4.14 WTV development for 99 vertices wing at 0° AoA

but dies down when it reaches the end of the upstroke phase. The wing-root vortex is also observed to be weak due to the low rotational motion of the wing root.

The Fig. 4.15 shows the visualisation result for the 99 vertices wing at all AoA during the 0.25 T time step which is during the down-stroke motion. As the AoA of attack increases, the results show that the strength of the vortex will also increase. However, it is important to note that not only the strength of the WTV increases but also the wing-root vortex will also increase. The presence of the wing-root vortex can initially be seen at 15° AoA where the wing-root vortex increases with increase as the wing moves downward. However, at 15° AoA, the wing-root vortex is still relatively small. But at 30° and 45° AoA, the wing-root vortex increases in strength dramatically to a point that the wing-root vortex has the same strength if not more strength than the WTV. The reason why the wing-root vortex increases in strength is because the bottom of the wing faces more and more to the incoming freestream

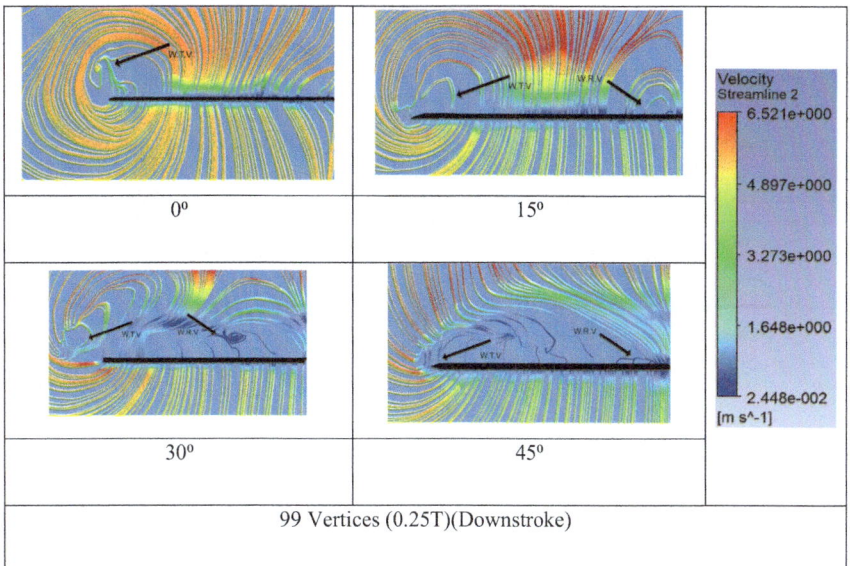

Fig. 4.15 WTV development for 99 vertices wing downstroke motion

which makes spillover effect to not only rely on the motion of the wing to happen but is also caused by the incoming airflow.

The Fig. 4.16 shows the visualisation result for the 99 vertices wing at 0.75 T which is during the upstroke phase. The typical and simple development of the wingtip and wing-root vortices during the down-stroke phase can be seen in the 0° AoA where the phase begins as the wing moves up and the WTV that was left during the down-stroke phase begins to die down. As the wing continues to move upwards, the motion will push the air upwards and airflow begins to spill over from the top of the wing surface and into the bottom of the wing surface. This then causes a similar WTV, but the direction of the vortex is the opposite. Also, the WTV grows as the upstroke motion continues but dies down when it reaches the end of the upstroke phase. The wing-root vortex is also observed to be weak due to the low rotational motion of the wing root.

However, a different observation can be made when the AoA of the wing is at 15°. Because the bottom surface of the wing faces the incoming airflow, it can be observed that the incoming airflow started to flow from the bottom of the wing and spill over to the top of the wing. This cancels out the WTV that is generated by the upstroke motion of the wing. Plus, due to the same incoming flow that is hitting the bottom of the wing, the wing-root vortex also can be observed. However, since the AoA is still low, the initial WTV that moves from the top of the wing to the bottom of the wing is still observable and the wing-root vortex is still just as strong as the WTV.

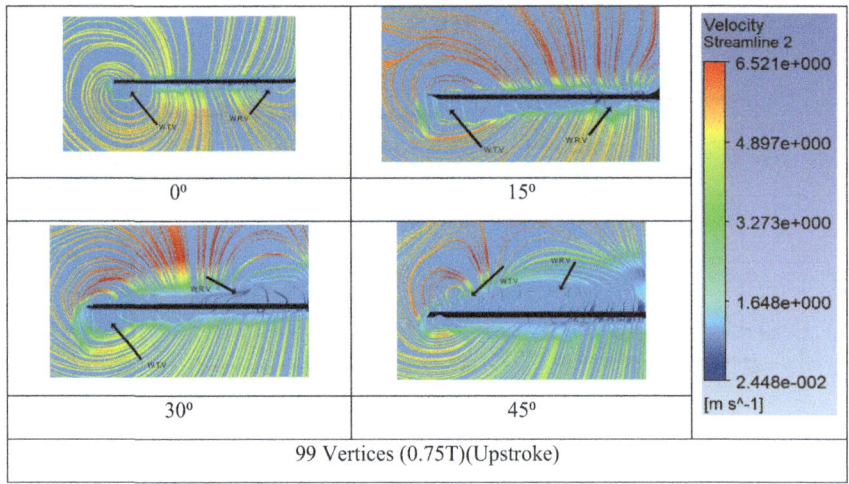

Fig. 4.16 WTV development for 99 vertices wing upstroke motion

As the AoA continues to increase, the power of the vortex generated due to the incoming airflow increases and, at 30° and 45° AoA, there are little to no trace of the wingtip or wing-root vortex that moved from top to bottom, the only bottom to the top. Plus, with increasing AoA, also comes with the increasing wing-root vortex. Since the WTV is diminished by the counteracting vortices, the wing-root vortex is now dominant where at certain instances, the WTV is completely overpowered by the WRV.

From the results, it can be seen that the overall development of the wingtip and wing-root vortices with changing AoA and changing time step is similar for all wing geometries. The vortices developments are the same because the vortices are dependent on the motion of the wingtip and, since the wingtip motion of each wing geometry is the same, there is little difference in the pattern of the development of the vortices. There, the difference, however, can be seen in the magnitude of the vortex with the 5 vertices wing generating the highest local wind velocity followed by the 21 vertices wing, the 29 vertices wing, the 49 vertices wing, and the 99 vertices wing. This is because of the location of the curves at the trailing edge. With straighter trailing edge, there is more wing surface that pushes the wing and thus generating larger local wind velocities.

The question that remains now is how wingtip and wing-root vortices affect the lift and drag generation. In terms of lift, it can be observed that the strength of the wingtip and wing-root vortices correlates directly with increasing lift regardless of which vortex is more dominant. However, it can be also observed that the direction of the vortices also determines the upward and downward force where vortices that move from the lower surface of the wing to the top surface of the wing generate a positive lift while vortices that move from the top of the wing to the bottom of the wing generates a negative lift (or force). This can be seen in the visualisation results for 0° AoA for any wing geometries. The time-averaged lift generated by the wing at 0° AoA is partially due to the low angle that generates low lift but also due to the exceptionally low lift generation during the wing upstroke phase. During the upstroke phase, the lift time-averaged lift generation is almost zero. This is because the wing generates a negative (or downward force) but is countered by the lift generated by the LEV that was discussed before. The same observation can also be made for the 15° AoA but, since the vortex that moves from the top of the wing to the bottom of the wing has been counteracted, the distinction is less clear.

In terms of drag, the strength of vortices has also shown to correlate directly with the generated drag regardless of the direction of the vortices. However, it can be observed that the presence of the wing-root vortex has a large impact on the generated drag. This can be seen with the sudden jump in generated drag as the AoA increases beyond 15° AoA. This correlated directly with the larger presence of the wing-root vortex at AoA 30° and 45°. While this does not mean that either WTV or wing-root vortex play a bigger or larger role in drag generation, it does show that the addition of both increases in strength with increasing AoA causes the increase in drag generation.

The wing geometry itself is a unique wing shape since the shape was generated from a specific batwing using a novel method. Therefore, there are no previous works that can be used as a validation point for the generated lift, drag, and lift over drag values. However, since the wings themselves are based on natural wings, several observations, and phenomenon that is observed in this book can be validated by observations found in natural wings.

Reference

G. Sachs, Aerodynamic cost of flapping. J. Bionic Eng. **12**(1), 61–69 (2015). https://doi.org/10.1016/S1672-6529(14)60100-1

Chapter 5
Final Thoughts

The main goal of the book was to study the aerodynamic effects of a bio-inspiration approach to MAV design. This was done by designing a flapping-wing MAV design that was derived from observations made on batwings. From the works that have been done, several final thoughts can be made.

The first conclusion that can be made is regarding the bio-inspiration design approach. In terms of wing geometry mimicry and simplifications, two types of wings can be produced: close to nature and close to generic wings. The features that changed mainly the trailing edge of the wing. In terms of wing motion, the wing motion of the wing model was simplified to only include the flapping and bending motion of a batwing by adding a passive flexible mechanism to the wing. The final result of the wing form mimicry and simplification process was a series of designs that have different trailing edges and different wingtip motions all of which are based on or derived from observations found in a natural wing but are far simpler and mechanically feasible.

The second conclusion that can be made has to do with the aerodynamic implications of the batwing mimicry and simplifications. In terms of the effects of wing geometry, the results have shown that there is a trade-off between the two types of wings where wings that are closer to natural wings have shown to be lower lift, lower drag, and delayed stall angle. This observation is mostly true at later AoA with the 99 vertices wing having a 0.12–7.3% lift over drag improvement over the conventional 5 vertices wing. The LEV plays a role in both lifts with larger and stronger the LEV, the greater the generated lift.

The third conclusion that can be made is in terms of the method of designing a flapping wing for a flying MAV. From the results that were obtained, it was determined that the best wing design among all the wings tested in this book is the 99 vertices wing with a low spring stiffness flexible mechanism because it is this wing design combination that has shown to produce the best lift over drag value over wide types of conditions. However, more important final thoughts that can be made lies in the approach for the design process. It can be concluded that the wing geometry with the lowest drag should be used and the flexible mechanism will play a role in enhancing the lift. The result is a wing design with the best lift to drag ratio. In terms of application, the wing is best as a hand-launched MAV with a hanging wing configuration that lands by gliding down.

The main goal of this book is to contribute to the design and final production of a bat-inspired flapping-wing MAV. While steps have been made in the right direction, there is more work that needed to be done. One of the areas that still needed to be explored is the flight conditions that this book did not touch since it is outside of the scope of the book. For example, the different wind speeds and by extension different flight regimes needed to be explored since there is yet a gap of knowledge in terms of how the wing design will operate under different flight regimes.

Another future work that needed to be done is in terms of the wing geometry itself. The works done have exposed several important parameters that can be manipulated and optimised. Parameters such as the size and location of the trailing edge curve, and the location of the pivot for the flexible mechanism. Also, several other wing geometries can be derived from the same method that has not been tested since the book is the constraint to be able to test only a handful number of wings. Wings that are drawn using a different number of vertices can be studied and perhaps a better wing design can be found. In the same vein of the wing derivation, other possible future works are using the same method but done on a different species of bat or perhaps maybe even on different natural wings such as birds or bats to test the robustness of the method.

In terms of understanding the aerodynamics of a flapping-wing system, several phenomena needed to be explored and are not explored in this book. Aerodynamic phenomena such as the effect of the body of the MAV where, study assumes that the wing was free-floating, and the body of the MAV was not taken into consideration. The result was the wing-root vortex appears to play a bigger role which should not have been there if the MAV body was taken into consideration. Another phenomenon that was ignores is the wing-to-wing interaction. In the simulation work (which was vital in visualising the airflow), a half model was used. While assumption was good enough for the book, how the generated vortices and wake structure would have formed if the second wing would have been there been also not explored.

Finally, a future recommendation that can be made concerns the development of the flying MAV itself. The work that has been done is mainly done in a laboratory condition and does not yet be able to produce a flying device. Several considerations of MAV design have been ignored such as power requirements, stability, and control system. Power requirements are ignored since the wing inertia requirements are not studied. Also, since the lift generation is measured as a time-averaged lift, the overall stability of the system is not examined. And since the wing is tested only in forward flight and not manoeuvre flight, the flight control is not considered. All these issues needed to be addressed before a flying MAV can be designed. In the same vein of the production of a flying MAV, there are still works that have been done that showcase the full potential of nature to machine bio-inspired design approach. Nature to the machine means that no work has yet to be done that has taken the observations done by an individual natural wing and transformed it to a flying device that not only

5 Final Thoughts

uncovers new parameters and vital design approaches for producing a flying MAV but also uncovers new observations that can be made further the understanding of the natural world.

Index

A
Active morphing, 29
Aerodynamics, 1, 4, 21, 28–30, 32, 74
Angle of attack, 5, 12, 13, 29–31, 43, 45, 50
ANSYS, 49
Approach flight, 10

B
Bat wing, 8, 15, 32
Bending, 13, 14, 21, 22, 31, 32, 34, 40, 57, 73
Bio-inspiration, 1–3, 7, 8, 14, 15, 28, 39, 41–43, 73
Bio-mimicry, 17
Bird wing, 3, 32
Blending function, 51
Butterworth filter, 45

C
Cambering, 13, 18, 21, 22, 31
Capture flight, 10
CATIA, 41, 45
Chiroptera, 5, 8, 10
Computational Fluid Dynamics (CFD), 21, 25, 57
Computer Aided Design (CAD), 15, 41, 42, 45
Conventional wing, 16, 17, 29, 55, 57, 63
Cynopterys brachyotis, 5, 10, 41, 52

D
Dactylopatagium, 9
Data Acquisition System (DAQ), 45

E
Ecological factors, 12
Elliptical wing, 16, 43, 55, 65

F
Finite Element Analysis (FEA), 50
Fixed wing, 1, 2
Flapper, 40, 45–47, 57
Flapping frequency, 3, 5, 12–14, 23, 29, 43, 45, 46
Flapping wing, 2–4, 7, 29, 32, 39, 40, 43, 47, 53, 55, 73, 74
Flapping Wing Micro Air Vehicles (FW-MAV), 2–4, 51, 52
Flex angle, 14, 29
Flexion, 12, 14
Flight behaviour, 7, 8, 10–12
Flight kinematics, 7, 12, 45
Flow field, 39
Fluent, 49, 50
Fluid Deposition Modelling (FDM), 46, 47
Fluid Filament Fabrication (FFF), 46
Fluid-Structure Interactions (FSI), 49
Folding, 13, 14, 21, 22, 31, 32, 53
Foraging, 10

G
General wing shape, 16
Generic wing, 16–19, 30, 32, 33, 42, 54–57, 65, 73

Geometry generation, 40, 41, 65
Global value, 45, 52, 54, 57, 63, 65
G. Soricina, 10

I
Insect wing, 3, 32
In vivo, 14, 21, 54

K
K-epsilon, 50

L
L. Yerbabuenae, 10

M
Manoeuvrability, 2, 3, 9–11
Mechanical noise, 47, 57
Meshing, 50

N
Navier stokes, 50
Negative mesh, 50
Non-dimensional, 49

P
Passive morphing, 29
Pipistrelle, 10
Plagiopatagium, 9
Poly-Lactic Acid (PLA), 47
Prey retreival, 10
Propatagium, 9
Pursuit manoeuvre, 10, 11

R
Rigid body, 57

Roosting, 11
Root to tip, 42
Rotary wing, 1, 2, 29

S
Skeletal structure, 16
Stall angle, 29–31, 43, 54, 55, 57, 61, 63, 65, 73
Strain guage, 45
Streamline, 60, 65
Sweep angle, 14, 29
Synectic, 2
System coupling, 49

T
T. Brosilliensis, 10
3D print, 45, 46
3d realistic, 15, 16
Time step, 63, 68
Transient structure, 50
Twisting, 13, 21, 22, 31–33
2d wing, 18, 20, 21

U
UAV, 51
Uropatagium, 9

W
Wing area change, 13, 14
Wing beat, 29
Wing morphology, 3, 8

Z
Zeroing, 57, 71

The manufacturer's authorised representative in the EU is Springer Nature Customer Service Centre GmbH, Europaplatz 3, 69115 Heidelberg, Germany. If you have any concerns regarding our products, please contact ProductSafety@springernature.com

Printed and bound by CPI Group (UK) Ltd, Croydon, CR0 4YY

25/03/2026

02078193-0017